Lecture Notes in Artificial Inte

Subseries of Lecture Notes in Computer Science
Edited by J. Siekmann

Lecture Notes in Computer Science

Edited by G. Goos and J. Hartmanis

Andranick S. Tanguiane

Artificial Perception
and Music Recognition

Springer-Verlag

Berlin Heidelberg New York
London Paris Tokyo
Hong Kong Barcelona
Budapest

Series Editor

Jörg Siekmann
University of Saarland
German Research Center for Artificial Intelligence (DFKI)
Stuhlsatzenhausweg 3
D-66123 Saarbrücken, Germany

Author

Andranick Tanguiane
Fernuniversität Hagen
Postfach 940, D-58048 Hagen, Germany

CR Subject Classification (1991): I.2, I.5, E.2

ISBN 3-540-57394-1 Springer-Verlag Berlin Heidelberg New York
ISBN 0-387-57394-1 Springer-Verlag New York Berlin Heidelberg

Typesetting: Camera ready by author
Printing and binding: Druckhaus Beltz, Hemsbach/Bergstr.
45/3140-543210 - Printed on acid-free paper

Foreword

In this book I summarize my studies in music recognition aimed at developing a computer system for automatic notation of performed music. The performance of such a system is supposed to be similar to that of speech recognition systems: acoustical data at the input and music score printing at the output.

In this essay I develop an approach to pattern recognition which is entitled *artificial perception*. It is based on self-organizing input data in order to segregate patterns before their identification by artificial intelligence methods. The performance of the related model is similar to distinguishing objects in abstract painting without their explicit recognition.

In this approach I try to follow nature rather than to invent a new technical device. The model incorporates the *correlativity of perception*, based on two fundamental perception principles, the grouping principle and the simplicity principle, in a very tight interaction.

The grouping principle is understood as the capacity to discover similar configurations of stimuli and to form high-level configurations from them. This is equivalent to describing information in terms of generative elements and their transformations.

The simplicity principle is modeled by finding the least complex representations of data that are possible. The complexity of data is understood in the sense of Kolmogorov, i.e., as the amount of memory storage required for the data representation.

The tight interdependence between these two principles corresponds to finding generative elements and their transformations with regard to the complexity of the total representation of data. This interdependence justifies the term "correlativity", which is more than relativity of perception.

The model of correlative perception is applied to voice separation (chord recognition) and rhythm/tempo tracking.

Chord spectra are described in terms of generative spectra and their transformations. The generative spectrum corresponds to a tone spectrum which is repeated several times in the chord spectrum. The transformations of the generative spectrum are its translations along the $\log_2$-scaled frequency axis. These translations correspond to intervals between the chord tones. Therefore, a chord is understood as an acoustical contour drawn by a tone spectral pattern in the frequency domain.

Time events are also described in terms of generative rhythmic patterns. A series of time events is represented as a repetition of a few rhythmic patterns which are distorted by music elaboration and tempo fluctuations associated with the tempo curve. The interdependence between tempo and rhythm is overcome by minimizing the total complexity of representation, e.g., the total amount of memory needed for storing rhythmic patterns and the tempo curve.

The model also explains the function of interval hearing, certain statements of music theory, and some phenomena in rhythm perception.

Generally speaking, I investigate hierarchical representations of data. In particular, I pose the following questions:

(a) Why a hierarchy?

(b) Which hierarchy? and

(c) How does the hierarchy correspond to the reality?

From the standpoint of the model, the answers to these questions are, respectively:

(a) A hierarchy makes a data representation compact, which is desirable in most cases;

(b) consequently, a better hierarchy is one which requires less memory for the related data representation; and

(c) under certain assumptions such a hierarchy reveals perception patterns and causal relationships in their generation, making the first step towards a semantical description of the data.

One can see that the main distinction of this approach is finding optimal representations of data instead of directly recognizing patterns. In a sense, analysis of patterns is replaced by synthesis of data representations. Since self-organization is used instead of learning, the threshold criteria used in most pattern recognition models are avoided.

The correspondence between music perception and the performance of the model, together with the diversity of its applications, can hardly be regarded as simply a coincidence. It makes an impression that the model really simulates certain perception mechanisms. Probably, the related model can be applied to speech recognition, computer vision, and even simulation of abstract thinking. All of this is a subject for discussion.

This book has been written during my work in Grenoble at the ACROE–LIFIA (Association pour la Création et la Recherche sur les Outils d'Expression–Laboratoire d'Informatique Fondamentale et d'Intelligence Artificielle).

I acknowledge Professor Philippe Jorrand, the director of the LIFIA, and Dr. Claude Cadoz and Dr. Annie Luciani, the heads of the ACROE, for inviting me to Grenoble and the stimulating working conditions; my colleagues, especially Martial Barraco, for their friendly attitude and help in various domains; and Remi Ozoux, a student of ENSIMAG (Ecole Nationale Supérieure d'Informatique et de Mathématiques Appliquées de Grenoble), who has fruitfully collaborated on the project and who has programmed the latest version of the algorithm for chord recognition.

Hagen, 14 September 1993   Andranik Tangian (Tanguiane)

Contents

List of Figures

List of Tables

Chapter 1

Introduction

1.1 Formulation of the Problem

By analogy with speech recognition, by *music recognition* we understand studies in computer simulation of music perception which contribute to developing systems for automatic notation of performed music. The performance of a system for automatic notation is supposed to be analogous to that of speech recognition systems: Acoustical data at the input and music score printing at the output.

Music recognition is needed in:

- *Music education* (computer listening to pupils and judging their skills, automatic accompaniment);

- *composition* (automatic music printing);

- *computer-aided performance* (simulating ensemble interaction of the computer and the soloist);

- *music publishing* (acoustical input of musical data);

- *musicology* (automatically transcribing and analyzing live performance, in particular, folk music; facilities for music data input for music data bases);

- *recording engeneering* (visualizing recording and tape editing).

Music recognition systems are also desirable for music lovers and amateur bands who would like to imitate their favorite performers. Music recognition devices could meet the demand for precise transcriptions of musical pieces which are not available from music publishers.

A complete system of music recognition consists of three interfaced subsystems for:

- *Acoustical recognition,*

- *music analysis,* and

- *music printing.*

The task for acoustical recognition is to determine the number of simultaneously sounding parts along with their dynamical and timbral specifications, to recognize instruments, to segment the signal into locally constant and transient segments, to determine periodicity, and to derive pitch trajectories for all voices. The output of this stage of analysis should resemble MIDI messages.

Music analysis includes recognition of time, tempo, tonality, note values and their relative durations, techniques of execution, dynamics, and other musical characteristics which are fixed by notation. At the given stage of data processing the acoustical information is interpreted semantically and represented symbolically.

Music printing, i.e. printing of musical scores, is rather a technical problem which has been solved already. A detailed survey of systems for music printing can be found in (Hewlett & Selfridge-Field 1990; 1991).

Most of the problems arising in music recognition have been considered in different disciplines. For example, pitch and timbre recognition are studied in psychoacoustics, more technical items as signal segmentation are developed in speech recognition, tonality determination is discussed in quantitative musicology, etc. However, there are two specific items inherent in music recognition:

- *Processing compound acoustical signals,* and

- *interpreting acoustical events in terms of elements of musical composition.*

Various particular problems arise both in audio data processing and symbolic coding of semantical information which is carried by the musical signal.

Thus developing a system for automatic notation of performed music requires solving new problems and implementing the models developed in other disciplines. On the other hand, the importance of music recognition for music and musicians is difficult to overestimate. It is similar to that of speech recognition for computer users and computer business. All of this makes music recognition to be a great challenge for computer scientists.

1.2   Brief Survey of Music Recognition

The seek for automatic recognition of music goes back to Bartok and Sieger who have realized its necessity in ethnomusicology studies. The latter have even used a device called "melograph" which could draw the fundamental frequency curve of a melody (Schloss 1985).

The works in computer recognition of music date back to early 70ies. The first experiments on automatic notation of monophonic music are realized by Sundberg & Tjernlund (1970) and Askenfelt (1976). According to Piszczalsky & Galler (1977) and Roads (1980), Ashton (1971) and Knowlton (1971; 1972) developed a program for transcribing polyphonic music performed on a computer-wired keyboard, having prepared grounds for some later commercially available devices (Roads 1982; 1987; Wyse Carl Disher & Labriola 1985).

The first integral computer system for music recognition was developed by Moorer (1975; 1977) and was, notably, oriented towards polyphony, although its capabilities were severely limited. The music which was to be analyzed was assumed to have two parts, only pitched sounds were admitted (bell-like, or percussive voices were prohibited), certain pitch combinations were avoided (primes, octaves, twelves, etc.), rhythmic structure had to be simple and the tempo had to be constant. In short, only specially selected and specially performed music could be recognized. Nevertheless, Moorer's system proved the realizability of automatic notation and became the starting point for a series of succeeding works.

Considerable progress was achieved in recognition of monophonic music (Piszczalsky & Galler 1977; Piszczalsky Galler Bossemeyer Hatamian & Looft 1981), especially owing to the application of artificial intelligence methods (Chafe Mont-Reynaud & Rush 1982; Foster Schloss & Rockmore 1982; Mont-Reynaud 1985; Mont-Reynaud & Goldstein 1985). It was proposed to use the knowledge about music of a particular style in order to describe all possible combinations of musical elements, among which the recognition system had to choose that combination which most closely resembled the input data. Some comments follow on the strong and weak aspects of this approach.

The use of artificial intelligence methods made it possible to juxtapose rhythm and pitch information so as to take into account rules of elaboration, and, by comparison to earlier works, to widen the admissible variety of recognizable rhythmic figures and melodic passages. A quite complex melody from the end of Mozart piano sonata K.333 performed without any restrictions was recognized in (Chafe et al. 1982). In this case the peculiarities of musical style were taken into account for correcting errors of acoustical recognition.

However, alongside evident advantages, the methods of artificial intelligence exclude applications of the same model to music of different cultural traditions, since stylistic peculiarities which are not anticipated in the model are usually misinterpreted. In other words, general capabilities of music perception are not adequately reflected in the model which is endowed only with particular knowledge.

Another weak point is the impossibility of adapting the model for running in real time, since only passages of considerable length can be analyzed, a handicap which results from the use of statistical hypotheses whose reliability is dependent on the amount of data processed. That is quite different from human perception; indeed, for example, rhythm is easily recognizable at the very first measures.

Notable among studies concerned with automatic notation of monophony is Niihara & Inokuchi's (1986) system for recognizing Japanese folk singing. A singing voice, being less stable than an instrumental sound, is much more difficult for acoustical analysis. The orientation towards folk music, of practical interest to ethnomusicologists, also distinguishes this study from purely laboratory experiments. However, the folk tunes used in the recognition experiments are too simple for judging the merits of the system.

By about 1985, ten years after Moorer's pioneering work, papers appeared which were devoted to some special topics in music recognition. The problems considered in these publications can be classified into the following two branches:

- *Chord recognition, or voice separation,* and

- *rhythm/tempo recognition.*

In papers on voice separation and note identification (Schloss 1985; Chafe Jaffe Kashima Mont-Reynaud & Smith 1985; Chafe & Jaffe 1986) account is taken of rules of counterpoint, knowledge about the previous context, specifications of the voices used, and so on. It has been proposed also to separate voices by the detection of the asynchronism in tone onsets. This idea is implemented in a system for recognizing Afro-Cuban percussion music, where the asynchronism of tone onsets is quite clear in the sharp attacks in tone envelopes of percussion sounds (Schloss 1985). More traditional material (an excerpt from Frescobaldi) is considered by Chafe & Jaffe (1986) who analyze a piano performance of a two-part passage with contrasting rhythmic patterns of voices.

A few Japanese papers report on the recognition of polyphony by means of advanced signal processing techniques (Katayose Kato Imai & Inokuchi 1989; Katayose & Inokuchi 1989a, 1990). However, the more detailed article (Katayose & Inokuchi 1989b) shows that the system is not yet sufficiently reliable. In a simple chord accompaniment to a melody, for almost every recognized chord there are either missed or false notes.

An interesting idea about the analysis of polyphony stems from the observation that different sounds even of the same instrument show individual characteristics of fluctuations in amplitude and pitch (Vercoe & Cumming 1988; Ellis & Vercoe 1991; Mont-Reynaud & Mellinger 1989; Mellinger & Mont-Reynaud 1991). These authors propose to track such fluctuations in order to separate the spectrum of a multivoice signal into classes of partials united by common law of motion. The asynchronism of tone onsets and differential development of simultaneous tones afford a basis for voice separation with respect to their dissimilarity. Recently, Mont-Reynaud & Gresset (1990) proposed the use of image processing models for the analysis of multivoice spectra versus time.

Among the studies on rhythm recognition some are of special interest.

A "strategic" approach to rhythm recognition is proposed by Longuet-Higgins (1976; 1987) and Longuet-Higgins & Lee (1982; 1984). Rhythm recognition is understood as finding a strategy in "listening" to music: A hypothesis concerning the rhythmic structure is formulated by the very first events, then it is continuously confronted with the current data, being modified if necessary, and finally a hierarchical structure of the rhythm is developed. A further elaboration of this idea is reported by Povel & Essens (1985) and Desain (1992). The former of the cited works deals with modeling an internal clock which is activated by temporal patterns. The latter is based on the notion of expectancy applied to rhythm perception.

A noteworthy approach to simulation of rhythm perception is proposed by Bamberger (1980) and Rosenthal (1988; 1989; 1992). The rhythmic structure of a melody is separated into simplest patterns which are arranged into repeating segments in order to give the whole structure a certain symmetry. A point of interest is that rhythm is understood as a means of data organization.

An advance in rhythm recognition in application of neuron nets was made by Desain & Honing (1989) and Desain Honing & de Rijk (1989). They suggest that each time interval between tone onsets should be put into correspondence with a neuron whose activation level is proportional to the given time interval. By transferring activation to each other, neurons filter the rhythm and represent it without irregular deviations. The result is a stable state of the net with simple ratios of activation levels of adjacent neurons. This stable state of the net is interpreted as the filtered rhythm. The novel aspect of the model lies in the understanding of rhythm as a simplified conceptual description of a sequence of time relationships. A similar approach to rhythm is discussed by Clarke (1987).

Finally, we mention the development by Dannenberg & Mont-Reynaud (1987) of a system which automatically accompanies a soloist and provides tempo tracking in real time. Their program follows a jazz improvisation, taking into account both time and pitch relationships. A further development is reported by Allen & Dannenberg (1990) and Dannenberg & Bookstein (1991). In spite of restrictions and limited reliability, the program is remarkable as an attempt of rhythm recognition in real time from current data.

The studies cited show that there has been considerable progress in music recognition as well as in understanding its main difficulties. It is recognized that the direct technical approach is nearly exhausted and that new musicological theories of music perception are needed (Richard 1990; Widmer 1990, 1992).

Summing up what has been said, we conclude the following.

1. The problem of music recognition and automatic notation is solvable. However, there are still many difficulties to be overcome.

2. The direct technical approach of the first developments may be replaced by more refined methods. Recent studies in music recognition exhibit two specific problems, namely voice separation and rhythm/tempo tracking. Some original ideas and sophisticated techniques of signal processing are suggested to the end of solving these two problems.

3. Further progress in music recognition is constrained by the lack of clear understanding of the nature of music perception. In particular, it turns out that there are no explicit definitions of notes, chords, rhythm, and tempo, which complicates their recognition. Therefore, in order to move forward, it is desirable to understand the nature of these concepts and the associated mechanisms of music perception.

1.3   Brief Survey of Artificial Perception

In the previous section we have pointed out some principal music recognition difficulties and have shown that common pattern recognition methods are not sufficient to surmount them. Therefore, we consider the artificial perception approach which has been developed mostly in computer vision and which turns out to be useful in music recognition as well. First of all let us characterize artificial perception in general and make references to principal contributors.

We distinguish the following two stages in pattern recognition:

(a) *pattern segregation,* i.e. grouping data into messages;

(b) *pattern identification,* i.e. matching the segregated messages to known memory patterns.

For example, the first stage corresponds to distinguishing lines, spots, etc. in abstract painting, but their associating with common concepts is the task for the second stage.

Some pattern recognition methods are based on direct identification of known objects in data streams without any intermediate processing. Such direct matching of objects to memory patterns can be quite efficient in certain applications (Freeman 1979).

Instead of matching to memory patterns, some objects can be directly identified by so called *invariant property methods,* i.e. by recognizing some invariant features of the objects which are common to all of their views (Pitts & Mc-Culloch 1947; Ullman 1990a). For example, biological cells can be recognized with respect to a "compactedness measure", which is defined to be the ratio between the cell's apparent area and its perimeter length squared. Cells that are almost round have a high score with respect to this measure, whereas long and narrow objects have a low score. It is important that such a measure is invariant with respect to rotations, translations, and scaling. Certain coefficients of the Fourier transform, object moments, and topological properties are some other examples of invariant characteristics useful for object recognition.

Another example of direct identification is given by the *alignment approach* (Ullman 1990a). This approach consists of two stages. In order to match an object to patterns stored in the memory, the transformations (rotations, translations, etc.) linking each pattern with the object are determined (this is said to be alignment), and then the memory pattern which is the closest to the object is chosen. It is important that the search of the pattern is performed over the patterns but not over the patterns under various transformations, since the transformations (alignments) are already known.

An example of applying similar ideas to character recognition can be found in (Neisser 1966; Preparata & Shamos 1985). In order to recognize a character, every character stored in the memory is matched to the given one, and the allowed transformation which provides the best correspondence between the memory pattern and the given character is to be determined. At the next stage, the memory pattern which provides the best correspondence is chosen.

However, in most practical cases some special techniques are used to segregate patterns before their identification. For this purpose a preliminary stage of data processing is added. The models used at this stage are said to be models of "not-intelligent", "not informed", "naive", or "early" perception. They are aimed at obtaining aggregate representations of input data by means of some description primitives to the end of object segregation, without their identification however. The identification, being based on confronting input data with some previously accumulated knowledge, belongs, strictly speaking, to the domain of artificial intelligence. In order to distinguish the models of "non-intelligent" perception from that of artificial intelligence, we shall refer to them as to *artificial perception* models.

The necessity of such artificial perception models was recognized by Zucker Rosenfeld & Davis (1975). It became clear that the "segmentation-and-label" problem was ill-defined, because the objects to be segregated depend on task and context (Marr 1982; Witkin & Tenenbaum 1983b).

The purpose of artificial perception models is recognizing structure in images. The importance of this task is explained by Witkin and Tonenbaum (1983b, pp. 482–483) as follows:

People's ability to perceive structure in images exists apart from the perception of tri-dimensionality and from the recognition of familiar objects. That is, we organize the data even when we have no idea what it is we are organizing. What is remarkable is the degree to which such naively perceived structure survives more or less intact once a semantic context is established: the naive observer often sees essentially the same things an expert does, the difference between naive and informed perception amounting to little more than labeling the perception primitives. It is almost as if the visual system has some basis for guessing *what* is important without knowing *why* ...

... The aim of perceptual organization is the discovery and description of spatio-temporal coherence and regularity. Because regular structural relationships are extremely unlikely to arise by the chance configuration of independent elements, such structure, when observed, almost certainly denotes some underlying unified cause or process. A description that decomposes the image into constituents that capture regularity or coherence therefore provides descriptive chunks that act as "semantic precursors," in the sense that they deserve or demand explanations.

However, recognizing structure is difficult because of the lack of its explicit definition. Therefore, most authors avoid giving strict definitions of structure, preferring to formulate the problem as finding regularity, symmetry, repetitiveness in the data, or, more generally, *perceptual grouping.*

The basic observation about grouping is that the perceptual system has a tendency to put together elements of the visual field in terms of "belongingness" (Palmer 1983, p. 287). Perceptual grouping was studied by Gestalt psychologists and explained as the capability to group elements with respect to their proximity, similarity in color, size, and orientation, and also continuity, closure, and symmetry (Wertheimer 1923). One of most important factors of grouping is the *"common fate"* of some elements which becomes apparent in dynamics when the elements move simultaneously in the same direction at the same rate.

The recognition of "common fate" is the underlying idea of the computer recognition of *shape from motion.* The related works started in 70ies were based on discovering the configurations of the elements which are united by similar displacements, i.e. on recognizing a configuration from the "common fate" of its elements. One can say that the trajectory of motion is recognized first, and the carrier of this trajectory is recognized as an object. Note that object recognition by motion doesn't require any special knowledge about the object, so that no learning is needed (Longuet-Higgins & Prazdny 1984; Shoham 1988).

This approach was applied to the analysis of two- and three- dimensional scenes, usually for rigid objects (Marr 1982; Gong & Buxton 1992; Meygret & Thonnat 1990; Navab & Zhang 1992; Thibadeau 1986; Ullman 1979, 1990b). Under certain constraints motion cues were used to recognize non-rigid objects as well (Terzopoulos Witkin & Kass 1988; Witkin Kass & Terzopoulos 1990; Pentland & Horowitz 1991).

A physically similar approach, but based on the analysis of optical flow instead of motion field was applied to the recognition of *shape from shading*. The only difference is that the motion field is a purely geometric concept without any ambiguity—it is the projection into the image of three-dimensional motion vectors, whereas the optical flow is a velocity field in the image transformation; hence it needs an additional constraint in order to uniquely determine the image transformation. The works on recovering shape from shading were started by Horn (1975). The related "variational approach" to optical flow was implemented into a computational model by Horn & Schunk (1981) and Ikeuchi & Schunk (1981) see Horn & Schunk (1993) and Ikeuchi (1993) for a retrospective.

The idea of "common fate" was also applied to modeling *textural grouping* in static images. It was notices that similar textural elements, in a sense united by a "common fate," can be grouped together, providing for a segmentation of the image. It was shown that the textural segmentation occurs as a result of differences in the first-order statistics of local features of textural elements rather than as a result of differences in the global second-order statistics of image points (Beck Prazdny & Rosenfeld 1983).

The texture was also used for recognizing local shape in images. The related problem is usually referred to as recognizing *shape from texture*. Gibson (1950) assumed that a homogeneous plane can be considered as covered with textural elements whose density is constant over the plane, whence the density gradient of the image is directly related, via perspective effects, to surface orientation. For example, the shape of a ball is easily perceived from the changes of the density of its ornamentation elements. Two different computer implementations of this idea were developed by Witkin (1981) and Kanatani & Chou (1989) which were generalized by Blake & Marinos (1990).

Another important problem of artificial perception is *recognition of contours* in images and recognition of *shape from contour*. Commonly contours are recognized by their continuity and by some characteristic local environment (change of color, texture density, etc.). It is noteworthy that contours are supposed to be determined by some kind of repetition of the same local environment. In a sense, this makes contours to be similar to trajectories, with the only difference that trajectories are drawn by some repetitious image element in time, whereas contours are drawn by some repetitious image element in space (Palmer 1983, p. 302). Some examples of contour recognition can be found in (Bolles & Cain 1983; Fischer & Bolles 1983; Lifshitz & Pizer 1990; Zhang & Faugeras 1990).

The approach to recognizing contours by representing them by simple primitives called "codons" is developed by Richards & Hoffmann (1976). These primitives are primarily image-based descriptors which have the power to capture important information about the three-dimensional world.

A similar approach is applied to modeling of a higher level of visual perception, corresponding to object recognition by recognizing their constituent parts (Ulmann 1990a). Similarly to a contour which is supposed to be generated by "codons", an object is supposed to be a combination of some primitives called "geons" like cylinders, boxes, etc. (Marr & Nishihara 1978; Brooks 1981). For example, a face contains the eyes, nose, etc., which can be recognized first in order to recognize a face. The number of primitives is supposed to be small (less than 50) and objects are typically composed of a small number of parts (less than 10) (Ullman 1990a).

The interaction between generic elements can be described by topological means and structural descriptions (Barlow 1972; Barlow Narasimmhan & Rosenfeld 1972; Milner 1974; Minsky & Papert 1988). Such methods use the idea of hierarchy, and this "feature hierarchy" is determined both by the constituent parts and by the way they interact with each other. Unlike other artificial perception techniques, these *object decomposition methods*, dealing with a high level of object description, are based on learning, since the image components and their interaction should satisfy some conditions. The review of recent publications on object description from contours can be found in (Mohan & Nevatia 1992; Ulupinar & Nevatia 1993; Bergevin & Levine 1993; Barrow & Tenenbaum 1993; Kanade 1993).

Among the works on modeling particular properties of visual perception, the recognition of *shape from stereo* should be mentioned. The related models process binocular images, using the geometric laws of perspective in order to recognize shape and three-dimensional motion of objects (Marr & Poggio 1976; Marr 1982; Kanatani 1984; Pridmore Mayhew & Frisby 1990). As in other models of artificial perception, no special knowledge about the objects is needed.

The enumerated models of particular visual functions (recognizing shape from motion, recognizing shape from shading, texture segmentation and recognizing shape from texture, recognizing contours and shape from contours, recognizing shape from stereo) are often used in combinations as modules in integral systems of computer vision (Horn 1986; Aloimonos & Shulman 1989). The complementarity of the perception cues implies a better performance of the integral systems of artificial perception.

It should be said that although most of the enumerated models of visual perception use certain physical constraints, they are based on the the assumption that the perception process is primarily data-driven. Such an assumption goes back to Gibson (1950, 1966, 1979) and even to Gestalt psychologists

(Wertheimer 1923) who have postulated the *preference for simple percepts* as a criterion for the *self-organization of perceptual data*. The simplicity principle was explained in terms of the tendency of self-regulating brain activity towards the minimum energy level consistent with the prevailing stimulation.

Later, the simplicity principle has been formulated in terms of information theory as the tendency towards the most economical description (Hochberger & McAlister 1953; Atteneave 1954; 1982). Leeuwenberg (1971; 1978) developed a coding theory where the the simplicity of a figure was estimated by means of the parametric complexity of the code required to generate it. Recently, the simplicity principle was formalized by using the notion of *data complexity* in the sense of Kolmogorov (1965) which is defined to be the amount of memory storage required for the algorithm of the data generation (Tanguiane 1990; Hoffmann 1992).

The simplicity principle is interconnected with the *hierarchization* in data representation. For example, the description of a dynamical scene in terms of objects and trajectories requires less memory than storing all the information about successive images which constitute the totality of data about the scene. At the same time, constructing such an economical representation of a dynamical scene results in the hierarchization of data description. At the lowest level, one has the initial data (stimuli, pixels). The patterns of the first level are similar configurations of data which are traced in successive images and recognized as moving objects. The relationships between corresponding low-level patterns in successive images determine high-level patterns which are associated with the object trajectories. Thus constructing an economical (simple) representation results in constructing a multi-level hierarchy of patterns, where the patterns of a lower level are carriers of the patterns of the higher level (Tanguiane 1990; Raynaut & Samuel 1992).

A general functional approach to understanding the hierarchical structurization in perception was proposed by Leyton (1986). He considered a multi-level architecture of data representation in terms of generative systems of analyzers and formulated structure postulates with respect to the grouping in the representations obtained. The recognition of structure was based on self-organization of data aimed at its economical description.

Thus the simplicity principle implies several important corollaries.

- First of all, it is a general criterion of quality of data representation.

- Next, the simplicity principle justifies hierarchical data representations as saving memory and provides the cues for finding optimal hierarchies for the data representation, corresponding to perception patterns (Witkin & Tenenbaum 1983a; 1983b).

- Moreover, the simplicity principle can help in overcoming the ambiguity in certain hierarchical representations discussed by Morita Kawashima

& Aoki (1992) and Moses & Ullman (1992).

- Therefore, the recognition problem can be formulated as an optimization problem in constructing data representations (Kass Witkin & Terzopoulos 1988; Friedland & Rosenfeld 1992).

- Finally, since any representation is already a description, finding optimal representations is the first step towards understanding the scene semantics (Rock 1983).

Thus we have briefly reviewed perception models of particular visual functions and models of self-organization of visual data. Perception models in audio data processing are developed mostly for the needs of recognition of speech in a noisy background and for music recognition (Bregman 1990; Darwin 1984; Handel 1989; Moore 1982; Warren 1982).

As in visual perception models, in audio pattern recognition two stages are considered, pattern segregation and pattern identification. For example, the first stage corresponds to distinguishing independent acoustical processes, like sounds from different sources or voices in polyphonic music, and the second stage corresponds to process identification, e.g. source recognition or melody recognition.

Until recently audio objects have been identified directly in audio data flows without any intermediate processing. This approach is quite sufficient for the recognition of speech of a single individual or for recognition of monophonic music.

Invariant property methods are used in speech recognition where phonemes are identified by their invariant characteristics (formants, i.e. typical spectral envelopes of vowels, the presence of high frequencies in certain consonants, etc.). In music recognition, tone patterns are recognized in chord spectra by their invariant harmonic structure (Moorer 1975, 1977; Chafe Jaffe Kashima Mont-Reynaud & Smith 1985; Chafe & Jaffe 1986; Katayose Kato Imai & Inokuchi 1989; Katayose & Inokuchi 1989a–b, 1990).

A kind of alignment approach is frequent in speech recognition where input phonemes and words are confronted to memory patterns. The alignment is also used in rhythm recognition under variable tempo (Chafe et al. 1982). As in vision, an input pattern is linked to all memory patterns by means of admissible transformations, and then the memory pattern which is the closest to the input pattern is chosen.

The need for the segregation of concurrent sounds (McAdams 1989, 1991a; Bregman 1990) poses a problem of modeling perceptual grouping in audio. In particular, the cited authors have considered the "common fate" principle for tracking simultaneous acoustical processes. The idea proposed is similar to that in visual scene analysis: The acoustical data are represented as a sequence of short-time spectral cuts, analogous to cinema frames in vision, and then one has to find spectral patterns whose partials synchronously develop in time, being united by a "common fate".

The related task in speech recognition is known as a "cocktail-party" problem where the overlapping phrases of different participants must be separated according to invariant features inherent in each voice. The voices are recognized by tracing the continuous development of certain spectral constituents. The corresponding models for voice recognition in polyphonic music are discussed by Vercoe & Cumming (1988); Ellis & Vercoe (1991); Mont-Reynaud & Mellinger (1989); Mellinger & Mont-Reynaud (1991), and with direct reference to visual analogy by Mont-Reynaud & Gresset (1990).

The recognition of audio structure by identity and by similarity is based on the same principles as that in visual scene analysis. Thus the recognition of audio structure by identity and similarity is proposed in studies on modeling rhythm perception (Clarke & Krumhansl 1990; Povel & Essens 1985; Rosenthal 1988, 1989, 1992). The rhythmic structure is recognized by finding similar rhythmic phrases and constructing hierarchical representations of time data based on repetitious segments.

Linking notes into melodies with respect to similarity of spectral patterns of voices is considered by Bregman (1990) and Bregman & McAdams (1979). In particular, in the cited works it is shown that notes are linked into a melody only if these notes are performed by the same voice, otherwise the perception of the melody becomes difficult, since a sequence of tones with different timbres is perceived rather as a timbral rhythm (cf. with timbral melody, *Klangfarben-melodie,* imagined by A.Schoenberg).

The methods of recognition of audio structure from intensity, loudness, and spectral density are reviewed by McAdams (1993). The related models are based on analysis of audio gradients which are similar to textural gradients proposed by Gibson. Similar models are used in visual processing in recognizing shape from texture.

The approach to audio localization and structurization from stereo is developed by Kendall & Martens (1984); Kendall Martens Freed Ludwig & Karstens (1986); Martens (1987) and Wightman & Kistler (1989). These authors developed models where sound localization and source segregation is performed with respect to interaural delay and spectral shaping introduced by head, pinna (outer ear), shoulders, and upper torso. The idea of recognizing an audio scene from comparing data from two distant sensors is similar to the idea of recognition of visual shape from stereo.

Even from our brief remarks one can conclude that audio and visual perception modeling have many common features. Both visual and audio scene analysis make use of "common fate" principle for modeling perceptual grouping. Scenes are recognized from motion, from local characteristics like intensity, textural, or spectral density, from finding identity and similarity, and from stereo. However, the computer approach to hearing is developed much less than the computer approach to vision. This point of view is shared by several authors, e.g. see McAdams (1991b).

Summing up what has been said in this section, we conclude the following:

1. The difference between artificial perception and artificial intelligence in pattern recognition is understood as follows. Artificial perception is used for discovering structure in visual and audio images by self-organization of data and segregation of patterns. Artificial intelligence is used for pattern identification by their matching to known concepts. Usually, the identification of already segregated patterns is much simpler than their recognition in data flows; thus artificial perception and artificial intelligence are complementary.

2. The artificial perception models of visual functions are classified into

    (a) recognizing shape from motion,

    (b) recognizing shape from shading,

    (c) texture segmentation and recognizing shape from texture,

    (d) recognizing contours and recognizing shape from contours,

    (e) recognizing shape from stereo.

    The artificial perception models of hearing functions are classified into

    (a) recognizing audio processes from voice motion,

    (b) recognizing audio structure from intensity and spectral density,

    (c) recognizing structure from similarity of audio events,

    (d) recognizing audio scene from stereo.

    These models do not use any particular knowledge on the patterns processed.

3. The "common fate" principle is considered as predominant in perceptual grouping. This principle is used directly in recognition of shape from object motion and in recognition of audio processes from voice motion. The segmentation of visual or audio image with respect to the "common fate" principle reveals the structure in the scenes analyzed. This means that this principle, being data-driven, contributes to the recognition of causality in visual and audio data.

4. The simplicity principle of perceptual grouping used in visual perception models is considered as a general criterion of data self-organization aimed at data reduction. The importance of this criterion is caused by its

Figure 1.1: Parallel primes, fifths, and octaves prohibited in counterpoint

generality and applicability to any data, independently of their type. In particular, such a criterion justifies the hierarchization principle of data organization, provides a means for chosing an optimal hierarchical representation and for overcoming the ambiguity in data grouping, and enables formulating the pattern recognition problem as an optimization task.

1.4 Development of Correlativity Principle

The development of artificial perception approach to music recognition was stimulated by music studies rather than by computer modeling. The first paper related to the subject was written when the author has studied orchestration with E. Denisov at the Moscow State Conservatory (Tanguiane 1977).

The starting point was a contradiction between some statements of music theory and musical practice. Namely, parallel voices (primes, octaves, and fifths shown in Fig. 1.1) are prohibited in the theory of counterpoint (Aldwell & Schachter 1978), while being widely used in orchestration and pipe organ mixture registers. Recall that a mixture register enables activating several pipes by a single key. Since these pipes are tuned according to a certain chord, playing a melody results in parallel leading of voices of the related chords.

In the available literature on orchestration and musical instruments this inconsistency of theory and practice is not explained. One can find some comments in H. Berlioz' *"Treatise on Orchestration"* (1855), where using mixture registers in pipe organs is severely criticized as "incompatible with the rules of counterpoint and unacceptable for musical ear."

However, according to the author's own experience in playing organ, mixture registers synthesize new timbres rather than provide a polyphonic effect. Moreover, every instrumental voice is always complex, being composed of a series of sinusoidal partial tones. Consequently, playing any musical instrument results in parallel leading of these partial tones, which is never considered as an undesirable harmonic effect.

Similarly, doubling bass or melody parts at unison or octave in orchestral arrangements makes the given voice brighter, not adding any harmonic quality. For the same purpose, a part can be multiplied at fifths, thirds, and other intervals, as in *Bolero* by M. Ravel (Fig. 1.2).

According to Tanguiane (1977), the prohibition against parallel voices in counterpoint and their use in orchestral arrangements is explained by the fact that voices in counterpoint and orchestration are not the same; to be precise, they have different musical meaning. A part in polyphony is more than simply a physical voice; it is a kind of melodic or harmonic *function.* This implies that a part in polyphony should be independent of other parts and well distinguishable. In orchestration, on the contrary, several instruments can be used to make an effect of a single line, contributing to the same compositional function. Therefore, the rules of counterpoint should be applied not to instrumental voices, but rather to the functional lines, simple or complex, corresponding to single instruments or groups of instruments, respectively, which depends on the context.

Thus there are two different cases in considering parallel voices. The first case arises while using several parallel voices as a single polyphonic part. Then these voices should fuse into one, synthesizing a new timbre, and for this purpose the use of parallel voices is acceptable. The second case arises when different parts are associated with different compositional functions. Then parallelisms should be avoided, since they result in the ambiguity in distinguishing these functions. One can conclude that after the voices have been grouped into functional lines, the rules of counterpoint should be applied to the entire groups (globally), but not inside the groups of voices (locally).

Having tested this conclusion in orchestral arrangements, the author proved that the results have corresponded to the expectations. Consequently, Berlioz' criticism against pipe organ mixture registers can be explained as caused by judging local effects from a global standpoint.

The observation that the perception tends to unite parallel voices into one and distinguishes well non-parallel voices is fundamental for the present study. It is an application of the principle of perceptual grouping of elements with a "common fate" to audio data (Bregman 1990).

At the beginning of the author's research in chord recognition the starting point was formulated as the following conjecture (see Tanguiane 1987; 1988a–b; 1989a–b).

Conjecture 1 (Unseparability of Parallel Voices) *Voices are not separable if they move in parallel with respect to the* $\log_2$-*scaled frequency axis.*

Thus if some partials move in the same direction at the same rate, these

Figure 1.2: The use of parallel voices in *Bolero* by M.Ravel

partials are united by a common law of motion. Conversely, if several groups of partials move differently, these groups can be distinguished by different laws of motion inherent in each group. Thus we obtain the following definition (Tanguiane 1987; 1988a–b).

Conjecture 2 (Part as an Acoustical Trajectory) *A part in polyphony, or a melodic line, is defined to be a dynamical acoustical trajectory drawn by a group of partials which move in parallel along the $\log_2$-scaled frequency axis in time. Such a stable group of partials is associated with the voice spectral pattern, or a note pattern.*

Thus recognizing chords is supposed to be realizable in dynamics, by voice tracking with respect to parallel motion of partials. The above definition of a polyphonic part prompts computational means for its recognition, similarly to that for the recognition of objects from motion (Ullman 1979; Marr 1982).

Conjecture 3 (Recognition of Parts by Correlation Analysis of Successive Spectra) *A polyphonic part which is drawn by translations of a voice spectral pattern along the $\log_2$-scaled frequency axis in time can be recognized by finding a stable subspectrum in successive short-time spectra of the musical signal. For this purpose one should perform correlation analysis of the successive short-time spectra with $\log_2$-scaled frequency axis.*

Next, it was noticed that if the same chord is repeated twice and all its voices have the same spectral pattern, the parts can be recognized not only as sustained, i.e. as determined by repeated notes, but also as linking any pair of notes, as if the voices were crossed. The above observation means that besides correlations between notes of adjacent chords, there should be correlations between notes of the same chord.

Thus note patterns correlate not only in dynamics, but also in statics. Since a melodic line is defined as an acoustical trajectory, a chord corresponds to a statical acoustical contour drawn by a note pattern. Such a similarity of melodic lines and chords is clearly seen when a chord is arpeggiated: Then the chord contour is spread out to a trajectory (Fig. 1.3). Conversely, a trajectory can be compressed into a contour, corresponding to a transformation of an arpeggio (melodic line) into a chord.

From the above reasons it follows that a chord can be defined and recognized in the same way as a melodic line. Thus we obtain the following conjectures (Tanguiane 1987; 1988a–b).

Conjecture 4 (Chord as an Acoustical Contour) *A chord is defined to be a statical acoustical contour drawn by a group of partials which is translated in parallel along the $\log_2$-scaled frequency axis. Such a stable group of partials is associated with a note pattern.*

Figure 1.3: Duality of chord contours and melody trajectories

Conjecture 5 (Recognition of Chords by Autocorrelation Analysis of Their Spectra) *A chord which results from translations of a note spectral pattern along $\log_2$-scaled frequency axis can be recognized by finding a stable subspectrum in its spectrum. For this purpose one should perform autocorrelation analysis of the short-time spectrum of the chord with $\log_2$-scaled frequency axis.*

One can easily see that the duality of contours and trajectories in audio is the same as that in vision (Palmer 1983). On the other hand, the approach to recognizing the structure of chords by finding similar constituents in their spectra meets Witkin & Tenenbaum's (1983b) idea that the visual structure should be based on identity of some elements.

Note that both a polyphonic part and a chord are definied in terms of translations of a voice spectral pattern with no reference to the concept of pitch. Therefore, these definitions are applicable to voices with no pitch salience, like bell-like sounds. Moreover, the separation of voices becomes possible without previous learning, since the task of voice separation is formulated not as recognizing some known subspectra but simply as finding repetitive subspectra.

In fact, we deal with a data representation based on finding *generative elements and their transformations*, or relationships between correlated patterns. Similar representations in vision were considered by Leyton (1986). Such a representation can be justified by the criterion of least complexity by Kolmogorov. Obviously, storing repetitive data as generated by a few elements is more efficient (requires less memory) than storing their totality.

The idea of representing data in terms of generative elements and their transformations was also applied to the problem of rhythm recognition and tempo tracking (Tanguiane 1991b; 1992a–b). The difficulty of the problem is caused by the fact that the tempo is perceived with respect to repeating rhythmic patterns, whereas the rhythmic patterns are recognized as repeated with respect to a certain tempo. It implies the ambiguity in interpreting each duration, since a duration can be identified either as given or as another value distorted by a tempo change.

In our model the interdependence between rhythm and tempo is overcome by application of the criterion of least complex representation of data. In case of representation of time events, some memory is used for storing generative rhythmic patterns, and some memory is used for describing their transformations, i.e. for storing their elaboration and tempo curve. The total complexity of the representation, i.e. the total amount of memory required, is the sum of these two amounts of memory. Therefore, we look for the representation of time data with minimal total complexity, which is shared between generative rhythmic patterns and a pattern of their transformations associated with the tempo curve.

Our approach to rhythm recognition can be formulated as the following conjecture.

Conjecture 6 (Rhythm Recognition by Optimal Representation of Time Data) *The task of rhythm recognition is formulated as finding the optimal (least complex) representation of time data in terms of generative rhythmic patterns and the pattern of their transformations associated with the tempo curve.*

Note that we use the same approach to three different problems, chord recognition, part recognition, and rhythm recognition. Its principal feature is that instead of finding some known patterns we construct representations of data based on recognizing repetitive messages and optimizing such a representation.

We think that the joint use of the two fundamental principles of perception, "common fate" principle and simplicity principle, is more than simply their union. The interaction of the two grouping mechanisms results in a new quality of perceptual grouping, making it much less ambiguous. At the same time, unambiguous grouping implies stable relationships between blocks of data, preparing ground for the hierarchization in grouping. Let us formulate this as the following conjecture (Tanguiane 1990).

Conjecture 7 (Principle of Correlativity of Perception) *By correlativity of perception we understand its capability to discover similar configurations of stimuli and to form high-level configurations from them. It is equivalent to describing the information in terms of generative elements and their transformations. Among all such representations of the data, the least complex representation must be chosen.*

Correlated blocks of data, said to be low-level patterns, determine high-level patterns which are formed by the relationships between the low-level patterns. This results in a natural hierarchy of patterns where similar patterns of a given level are carriers of the patterns of the next level.

The correlativity in its etymological meaning, "co-relativity," appears in the way patterns are chosen: Firstly, similar to each other (that is correlativity in the lower level); and, secondly, with respect to their interaction in the high-level patterns (that is correlativity between the two levels).

The criterion of least complexity provides the hierarchical arrangement of data with a feedback, guiding the process of data representation in the least complex way. This reduces the ambiguity in data representation, since usually there are alternative representations of the same data with different generative elements and different interactions between them. This way the idea of correlativity of perception is completed by the principle of optimality.

At present we are not capable to prove the hypothesis about the efficiency (optimality) of data representation based on generative elements and their transformations for a general case. Recently, the principle of correlativity was proved in application to perception of chords. Under certain assumptions it was shown that the optimal representation of a chord spectrum is the representation based on generative tones and their translations. Moreover, such a representation is unique, meaning the unique correspondence between optimal representation, chord generation, and perception of the chord (Tanguiane 1992c).

Thus assuming that the perception performs self-organization of data with respect to the criterion of their least complex representation, we explain why chords are perceived not as single complex sounds and not as collections of sinusoidal tones, but rather as sounds composed of sound complexes (notes). The unevidence of this perception phenomenon stems from the fact that the sensation of a chord is not caused by physical matters. Indeed, a sensation of a chord can arise from a sound produced by a single physical body, e.g. a loudspeaker membrane, or a piano board. If physical matters were predominant in sound perception, these sounds would be perceived either as entireties, or as collections of sinusoidal tones, which is not true.

It implies that the problem of chord recognition is a problem of *data representation* rather than that of recognition in a proper sense. Thus we come to the same conclusion as in case of rhythm recognition. In order to solve this recognition/representation problem, we characterize the related data representations and formulate the rules for their construction and quality criteria.

The following conjecture generalizes our approach to music recognition.

Conjecture 8 (Pattern Recognition as Optimal Representation of Data) *The problem of audio pattern recognition can be formulated as the representation of data which is based on their self-organization. The self-organization of data can be aimed at the data reduction and performed by constructing a hierarchy of generative patterns and their transformations. In a sense, analysis of patterns is replaced by synthesis of data representations.*

The above conjecture postulates the primacy of "pure" perception without any special intention to recognize something (cf. the quote from Witkin & Tenenbaum (1983) cited in the previous section). As a result, one obtains a hierarchical representation of data whose elements are treated as patterns (no matter meaningful or meaningless). Only then the patterns are identified (analyzed and matched to memory patterns), learned (stored in the memory for future matching), and labeled.

It is remarkable that representing the data in an optimal way, one can reveal the causality in data generation, implying its semantical interpretation. For example, a chord sound is originally generated by several sources of excitation, which can be recognized even if the chord sound is reproduced through a loudspeaker. We think that the optimality of physical interactions in nature should correspond to the optimality of their description.

Thus we formulate the last conjecture.

Conjecture 9 (Recognizing Physical Causality by Optimal Representation of Data) *Optimal representation of audio data reveals physical causality in the data generation. Thus optimal representation of data is a first step towards understanding their semantics.*

Similar reasons concerning recognizing causality in vision by simple representations are adduced in (Rock 1983, pp. 135, 335). He argues in favor of recognition of a common cause for co-occurring changes than the acceptance of coincidence.

To end this section, we recapitulate its main items.

1. The "common fate" principle is applied to the recognition of polyphonic voices. A polyphonic voice is understood to be a high-level pattern of acoustical trajectory which is drawn by translations of a generative voice spectral pattern in the frequency domain versus time. The recognition of voices is based on the same ideas as the recognition of objects from motion.

2. The "common fate" principle is applied to the recognition of chords. A chord is understood to be a high-level pattern of acoustical contour which is drawn by translations of a generative voice spectral pattern in the frequency domain. The recognition of chords is based on the same principles as the recognition of contours from texture.

3. The "common fate" principle together with the simplicity principle are applied to the recognition of rhythm and tempo. The task is understood as finding an optimal representation of time events in terms of generative rhythmic patterns and their transformations associated with the tempo curve. Using the Kolmogorov criterion of least complex representation

of data enables overcoming the interdependence between rhythm and tempo.

4. The interaction of the "common fate" and simplicity principles is formulated as the principle of correlativity of perception. The correlativity of perception is understood as its capability to discover similar configurations of stimuli (that is the correlativity between patterns of the same level) and to form high-level configurations from them. The ambiguity in data grouping is overcome by the use of Kolmogorov's criterion of least complex representation (this implies the correlativity between the levels of perception).

5. It is supposed that the audio pattern recognition problem can be formulated as the problem of constructing optimal representations of data. Such a representation contributes to revealing the underlying causality in the data analyzed.

1.5 Contribution to Music Recognition

The perception governed by stimulus relationships has been studied in psychology for a long time; for the review see, e.g., Chapter 8 in (Rock 1983). For example, stimulus relationships are predominant in perception of motion and in perception of orientation of objects in the environment.

We argue that stimulus relationships are also predominant in music perception. Indeed, one can see that musical information is transmitted not by sounds but rather by their relationships. In fact, a rhythm cannot be recognized by recognizing time events separately from each other. Another example is the recognizability of melodies in different keys implying the pitch to be less important than interval relationships between the tones.

Hence, we pose two fundamental questions:

* Which relationships and between which events are significant for music perception?

* How and why are they selected from all possible relationships?

Attempting to answer these two questions is our main contribution to music recognition. We suppose that the input data are represented hierarchically in the least complex way, and that such a representation itself reveals certain significant events and significant relationships between them.

Such an approach meets the Gestalt idea that objects are primary psychological representations (Posner 1978, Chapter 7; Posner & Henik 1983, p. 407). In the previous section we have mentioned that the problem of chord recognition as well as that of rhythm and tempo is the problem of data representation rather than that of recognition in a proper sense. (E.g. a chord sound reproduced through a loudspeaker cannot be identified with several sources other than by its special representation.)

Thus the problem of simulating music perception is formulated in terms of optimal data representation. It turns out that the relationships which are recognized by perception as significant coincide with the correlations of data which are used in constructing optimal data representations.

Generally speaking, subordinating music perception to the simplicity principle seems quite likely. It meets the idea of saving memory and of simplifying further data processing. In our study this *a priori* reason is theoretically and experimentally proved by establishing a correspondence between music perception and optimal representation of musical data.

The grouping with respect to two fundamental psychological principles, "common fate" principle and simplicity principle, constitutes a grouping mechanism which is said to be the correlativity of perception. We show that the two grouping principles control each other, reducing the ambiguity and resulting in a new quality of grouping. The similarity between the properties of music perception and that of the model based on the correlativity principle can be hardly regarded as simply a coincidence. It makes an impression that some general mechanism of music perception is discovered.

1.6   Summary of the Book

The presentation of the material is reversed with regard to the chronology of its development. The general concepts obtained recently are introduced first, whereas initial observations are given as applications. This is done for the same methodological reasons as in mathematics, where certain fundamental statements are formulated as axioms which result actually from successive generalizations.

In Chapter 2, "Correlativity of Perception", we formulate the principle of correlativity of perception and its mathematical model. The problem of recognizing generative elements and their transformations is formulated as finding correlated messages under various distortions of the scale. In order to perform a directional search for the scale distortions which provide high correlation of messages, a method of variable resolution is proposed.

In Chapter 3, "Substantiating the Model", we prove a series of mathematical propositions on representability of chord spectra as generated by spectral patterns of tones. It is shown that the representation of a discrete power spectrum of a chord as translations of a tone spectrum, which corresponds to the causality in the chord spectrum generation, is the least complex representation of spectral data. The demonstration is based on the deconvolution of chord spectra into the convolution product of irreducible spectra, similarly to the factorization of integers into primes.

In Chapter 4, "Implementing the Model", we show that instead of discrete power spectra it is reasonable to consider discrete Boolean spectra of chords (strings of 0 and 1). Such a simplification has two practical advantages: Computer processing is much more rapid, and Boolean spectra are much more stable with respect to distortions of spectral data, making the model better suitable for the analysis of real acoustical signal. In this chapter we formulate a theorem on the necessary condition for generative patterns in Boolean spectra of chords, and develop an algorithm for their finding.

In Chapter 5, "Experiments on Chord Recognition", the results of computer modeling are outlined. The algorithm proposed is tested and investigated on a series of experiments with synthesized spectra of chords. Possible recognition mistakes and performance of the algorithm are analyzed. The model shows not worse than 98%-reliability in recognizing four-part J.S. Bach polyphony for both harmonic and inharmonic synthesized voices. Besides, it is shown that the limits of the model's recognition capability are similar to that of trained musicians. If chords and voice spectral patterns are simple, they are recognizable by computer and by man. While chords and voices are getting more complex, man and machine fail to correctly recognize them almost simultaneously.

In Chapter 6, "Applications to Rhythm Recognition", the problem of tempo tracking and rhythm recognition is regarded from the standpoint of the principle of correlativity of perception. Repetitious rhythmic patterns (low-level patterns) are considered as carriers of time relationships which determine the perception of tempo (high-level pattern). In other words, rhythmic patterns are understood to be recognizable reference units for tempo tracking. The complexity of data representation is shared between the rhythmic patterns and tempo curve. The problem of rhythm/tempo recognition is formulated as finding the optimal (least complex) total representation of time data. In addition, the definitions of tempo, rhythm, and time are refined.

In Chapter 7, "Applications to Music Theory", we suggest an explanation of some properties of audio perception. In particular, we show that the logarithmic scale in pitch perception and the insensitivity of the ear to the phase of signal result in perceiving musical tones as entire sound objects but not as sound complexes. Besides, we propose a strict definition of interval with no reference to the pitch of tones and explain the function of interval hearing. We show that the properties of the auditory system mentioned, logarithmic pitch scale, insensitivity to the phase of signal, and interval hearing, provide the capability to track simultaneous acoustical processes and to recognize the causality in sound generation, which is necessary for the orientation in acoustical environment. In addition, some rules of music theory are justified as providing the conditions for adequate perception of music.

In Chapter 8, "General Discussion", some remarks on the further development of the artificial perception approach are made. In particular, we adduce reasons in favor of applying the artificial perception approach to arranging artificial intelligence data bases for knowledge representation. We suppose that optimal multi-level representation of knowledge, similar to optimal self-organization of sensory data which is studied in this essay, can find applications in computer vision and even in simulation of abstract thinking.

In "Conclusions" we recapitulate the main statements of the book.

Chapter 2

Correlativity of Perception

2.1 Introductory Remarks

In the present chapter we discuss a mechanism of perceptual grouping, the *correlativity of perception*, and suggest a way of its modeling. The background and successive steps in reasoning which have prepared the formulation of the principle of correlativity of perception are outlined in section 1.4; here we shall present it in detail.

In Section 2.2, "Principle of Correlativity of Perception", we illustrate the idea of correlative perception with examples of visual and audio perception. We also explain why the same data can be perceived in different contexts in a different way. From the standpoint of our model such a phenomenon is caused by the fact that the complexity of alternative representations depends on the context, implying the contextual dependence of the optimal representation which is associated with the percept. We also explain the effect of apparent motion from stroboscopic images.

In Section 2.3, "Model of Correlative Perception", the mathematical machinery for implementing correlative perception is outlined. We consider an example of motion recognition in a succession of instant images and show that the problem of representing the data in terms of generative elements and their transformations is reduced to the problem of finding correlated messages under various distortions of the data.

In Section 2.4, "Method of Variable Resolution", we describe how a directional search for generative elements and their transformations can be performed. The idea of the method is detecting the correlation between the images with reduced resolution wherein correlated elements are easy to find, and then gradually restoring the resolution while adjusting the images in order to maintain the correlation between the known elements. This way both similar elements and the transformations which provide for their high correlation are found.

In Section 2.5, "Complexity of Transformation as a Distance", we show that the complexity of transformation which links two patterns can be considered as a metrical distance between these patterns. An important property of this distance is that it formalizes the idea of likeness, not being based on the measure of identity. In particular, such a measure is efficient for recognizing similarity, e.g. when patterns have the same shape, being different in size.

In Section 2.6, "Distinctions of the Model", we discuss the peculiarities of our approach. We start with the questions which are to be answered by our model. Then we note that the hierarchization in data self-organization results from optimizing data representations and that the optimization criterion can replace threshold criteria in pattern segregation. We mention that pattern separation in the model is based on the pattern similarity but not on dissimilarity. Finally, we compare our artificial perception approach to pattern recognition with that of artificial intelligence.

In Section 2.7, "Summary of the Chapter", the main items of the chapter are recapitulated.

2.2   Principle of Correlativity of Perception

As already said in Section 1.4, by *correlativity of perception* we understand its capacity to discover similar configurations of stimuli, i.e. structurally arranged groups, and to form configurations of a higher level from them.

Configurations of stimuli themselves are called *low-level patterns*, and configurations of the relationships between low-level patterns are called *high-level patterns*.

This hierarchical scheme of data representation is endowed with a feedback, which guides the process of representing data with least complexity, where *complexity of data* is measured in the sense of Kolmogorov, i.e. as the amount of memory storage required for the algorithm of the data generation (Kolmogorov 1965; Calude 1988).

For example, in Fig. 2.1 one can see a collection of pixels (stimuli) which form symbols A (low-level patterns) which in turn form a contour of symbol B (high-level pattern). Obviously, instead of storing all the pixels, it is more efficient to store their configuration for one symbol A and then to store the contour of B.

Another important property of such a representation is that high-level patterns are stable with respect to changes of low-level patterns. For example, the substitution of Z's for A's in Fig. 2.1 would not influence on the perceptibility of B. The stability of high-level patterns with respect to changes of low-level patterns is described by Palmer (1982; 1983, p. 328).

Moreover, high-level patterns can be recognized without even recognizing

```
AAAAAAAAAAAAAAA
 AAAAAAAAAAAAAAAA
 AAAA           AAAA
 AAAA            AAAA
 AAAA            AAAA
 AAAA            AAAA
 AAAA            AAAA
 AAAAAAAAAAAAAAAA
 AAAAAAAAAAAAAAAA
 AAAA           AAAA
 AAAA            AAAA
 AAAA            AAAA
 AAAA            AAAA
 AAAA            AAAA
 AAAA            AAAA
 AAAAAAAAAAAAAAAAA
AAAAAAAAAAAAAAAAA
```

Figure 2.1: High-level pattern of *B* composed by low-level patterns of *A*

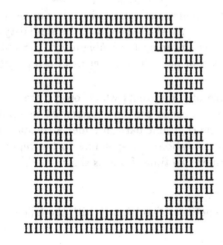

Figure 2.2: Pattern of *B* composed by unknown symbols

the underlying low-level patterns, Thus if some unknown, unrecognizable symbol were used instead of A, it still would be possible to recognize B by the relationships between these unknown symbols (Fig. 2.2).

This property is also inherent in human perception and learning. Most experiences are obtained from observing interactions of some objects whose peculiarities may be not so much important. For example, perceiving a traffic jam doesn't require any knowledge about the types of cars and their construction. Similarly, some ideas can be learned with no precise knowledge of basic concepts. For example, rules of arithmetics can be learned without strict axiomatization of numbers. Therefore, certain judgements can be made while considering relations between some concepts treated as "black boxes".

As we shall see further, recognizing intervals and chords doesn't require recognizing tones (pitch), which precisely corresponds to human perception. Indeed, most people lack absolute hearing (the capacity to recognize pitch of musical tones), but are capable to perceive intervals, melodies, and types of chords; e.g. distinguish between major and minor chords.

We restrict our attention to the case when important relationships arise between similar objects where the similarity is a cue for their recognition. In other words, we are looking for representations of data in terms of repetitive (correlating) messages, or generative elements, and their transformations.

At the same time, the recognition of similarity depends on some internal or external factors. In our model, such a factor is the total complexity of data representation which is determined by the complexity of low-level generative patterns and the complexity of the high-level patterns of their transformations.

Let us illustrate the influence of complexity criterion on recognizing similarity for further grouping. We shall show that the similarity can be relative, depending on the context, meaning the ambiguity of perception. Therefore, we explain the ambiguity of perception in terms of complexity of alternative data representations.

Consider a sequence of time events whose onsets are shown at the time axis in Fig. 2.3. This sequence can be represented in different ways. Its representation as a single rhythmic pattern under a constant tempo is shown in Fig. 2.4a. The representation corresponding to the repetition of the first three durations performed two times faster is shown in Fig. 2.4b where **R012** designates

- the call for the repetition algorithm **R** with the following parameters:

- begin from time **0**,

- repeat **1** time,

- perform the repetition **2** times faster.

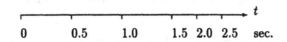

Figure 2.3: A succession of time events

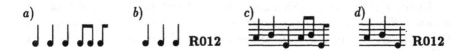

Figure 2.4: Four representations of the same succession of time events

Table 2.1: The Complexity of Representation of Time Events

	Represen- tation a)	Represen- tation b)	Represen- tation c)	Represen- tation d)
Complexity of rhythmic pattern	6 bytes	3 bytes	12 bytes	6 bytes
Complexity of its transformation	0 bytes	4 bytes	0 bytes	4 bytes
Total complexity	6 bytes	7 bytes	12 bytes	10 bytes

In this example the data representations has the following levels:

- Stimulus, corresponding to time events;

- Low-level rhythmic patterns which are groups of time events (either a single rhythmic pattern as in Fig. 2.4a, or two rhythmic patterns, the second being equal to the first but performed two times faster as in Fig. 2.4b);

- High-level pattern of the relationships between the rhythmic patterns (either trivial=no pattern in case of a single rhythmic pattern in Fig. 2.4a, or the pattern of repetition with a tempo change coded by the abbreviation **R012** in Fig. 2.4b).

Most likely, the given sequence of events is perceived as a long rhythmic pattern rather than a short one being repeated. This means that the representation in Fig. 2.4a is adequate, while that in Fig. 2.4b being inadequate.

However, if the same rhythmic sequence is placed into the melodic context shown in Fig. 2.4c, then the sensation of repetition becomes quite clear. Now the representation in Fig. 2.4d, where the idea of repetition is displayed, is rather natural and even can be considered as more adequate to perception than that in Fig. 2.4c.

To explain such an ambiguity in rhythm perception, that the same rhythmic passage can be perceived either as a long rhythmic pattern or as a repetition of a shorter one, estimate the complexity of the given representations. Suppose that one byte is needed to code a duration, and two bytes are needed to code a duration with pitch. Also suppose that to call the repetition algorithm, we need four bytes. Under such conventions the complexity of the given representations is estimated in Table 2.1. One can see that the representation of pure rhythm (Fig. 2.4a–b) as a long pattern is less*complex (6 bytes against 7), whereas the representation of the same rhythm in melodic context (Fig. 2.4c–d) is more efficient as a repeat (10 bytes against 12).

A similar effect of recognizing twice shorter or twice longer durations as a repetition of the same pattern with another time scale arises while listening to a fugue whose theme is played in augmentation or diminution. For example, such a device is used at the beginning of the fourth fugue from J.S.Bach's *The Art of Fugue* shown in Fig. 2.5 where T indicates the entry of the inversion of the theme, A indicates its entry in augmentation, and D indicates its entry in diminution.

Thus the perception correlativity appears in two instances: Firstly, the low-level patterns are chosen similarly to each other (that is correlativity at the same level); and, secondly, they are chosen with respect to their interaction in high-level patterns (that is correlativity between levels). This hierarchical scheme of data representation is provided with a feedback, the criterion of least

Figure 2.5: Theme in augmentation and diminution from Bach's *The Art of Fugue*

complexity. It guides the process of data representation in the least complex way, while the complexity being shared between the generative patterns and their transformations (see Table 2.1).

In a sense, the high-level pattern shows how the similarity of the low-level patterns should be understood in the given context, yet without such a high-level pattern the similarity of low-level patterns may be dubious, as in the example illustrated by Fig. 2.4a–b. This implies that the measure of similarity between low-level patterns is influenced by the way how they are confronted in the high-level configurations.

Threfore, we can speak of *contextual similarity*, or *functional similarity* with respect to some unifying high-level pattern. Within certain context two patterns may be perceived not so much dissimilar as if taken separately, since in the given context they are charged with a common function with respect to the high-level pattern.

Such a contextual similarity does arise in the representation in Fig. 2.4d, and does not when there are no additional cues as in case of melodic context in Fig. 2.4a–b. The effect of contextual similarity with respect to melodic cues arises also while recognizing a fugue's theme in diminution or in augmentation.

Revert to the example illustrated by Fig. 2.4a–b where three eights are not recognized as a replication of thee quarter notes. We have shown that in a melodic context these two groups of durations can be identified as similar. The same effect can be obtained with no melodic cues but by placing these two patterns in an appropriate rhythmic context.

For example, insert accelerating triads between the two patterns shown in Fig. 2.6a as it is shown in Fig. 2.6b. Owing to a gradual acceleration, the pattern marked by the bracket is unambiguously perceived as a repetition of the pattern formed by the first three durations. Here, the representation of the time events as a repeat of the first three events under a tempo acceleration is simpler than using complex fractional durations with which one can write down the same progression of events (see Fig. 2.6c). This means that the last three durations are recognized as similar to the first three durations, as required.

The stroboscopic effect, i.e. the illusion of apparent motion from successive images with slightly different locations of an object which is used in moving pictures and television, can be also explained within our model. Some psychologists explain the effect of apparent motion as the "solution to the problem of what is occurring in the world that might yield this unusual sequence of stimulation" (Rock 1983, p. 14 and Chapter 7). From our point of view, finding the meaning is not necessary, the motion illusion results from the least complex description of the scene in terms of generative elements and their transformations, i.e. objects and their trajectories.

For example, consider a sequence of movie frames with a flying ball in an

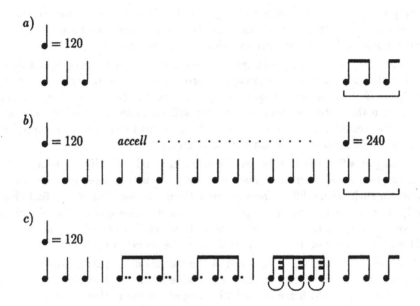

Figure 2.6: Contextual similarity of rhythmic patterns

invariable background shown in Fig. 2.7. From our standpoint, the apparent motion arises because of reducing the total visual information to its representation in terms of generative elements and their transformations (objects ant trajectories). Such an explanation doesn't use any assumption for "active perception", i.e. that "perceptual processing is guided by the effort or search to interpret proximal stimulus ... and the motivation for it must be the result of evolutionary adaptation" (Rock 1983, p. 16).

Note that the above effects of functional similarity, apparent motion, and ambiguity of perception, i.e. its dependence on the context, are explained with no reference to the meaning of percepts. We do not use any special constructions like conceptual frames (Minsky 1975) or meaningful settings (Palmer 1975). In other words, we provide for an explanation of percept dependence on the context which is *internal* with respect to perception, being based on no external cues (certainly, our explanation doesn't exclude others).

This pseudo-semantic self-organization of data is the most important feature of the correlative perception.

2.3 Model of Correlative Perception

Since we are looking for a representation of data in terms of variations of some generative elements, in other words, similar submessages, it is natural to apply methods of correlation analysis of data. However, since we are trying to recognize the transformations of generative elements, we have to use the correlation analysis under various deformations of data.

For example, to recognize a moving object in a cinema sequence as a more or less stable submessage, we can apply correlation analysis to distorted instant images. If these instant images are appropriately shifted, turned, stretched, etc., then the instant states of the object will correlate. In order to illustrate this idea, consider the problem of describing a dynamical scene at a computer display in terms of objects and their trajectories.

Imagine a flying ball in an invariable background. Identify the total visual information with a series of instant images, or with a cinematic sequence, wherein each frame differs from others only in the location of the ball (Fig. 2.7). In dynamics, a moving object is associated with a group of pixels which "move" according to a common law of motion (which have a "common fate"). This common law of motion is perceived as the object trajectory.

In order to separate the pattern of the ball from the pattern of the background, it is necessary to compare successive frames and to discover similar groups of pixels which are shifted with respect to each other. The pixels associated with the background are united by their immobility. Obviously, the groups of pixels associated with the ball are correlated in the statistical sense. If the ball's motion is complex and includes, for example, rotation, motion towards or away from the plane of the computer screen, etc., then the correlation model must provide for possible deformations of images, corresponding to the rules of motion in perspective.

Concerning the image transformations, it must be taken into account that every two images can be brought into correlation by some transformation. Hence there arises a risk of recognizing random configurations as intermediate states of the same object. Fortunately, such transformations are usually very complex. For this reason the model is provided with the criterion of least complexity, which rejects data descriptions which are too complex.

Thus, to describe a dynamical scene in terms of objects and their trajectories, it is necessary to analyze successive images and to discover pixel groups which are correlated under not very complex transformations. Next, among all such representations there must be found the one which is least complex. This scheme of calculations constitutes the *model of correlative perception*.

In statics, as contrasted with dynamics, the trajectory is replaced by the contour which is drawn by some generative element (as in Fig. 2.1). In general, everything said about trajectories is valid for contours with the only exception that correlation analysis is applied not to successive images but to the same image; in which case it is called autocorrelation analysis.

Note that the notion of contour is intended in a broad sense: A contour need not be continuous, closed, or even unidimensional. In this sense, the

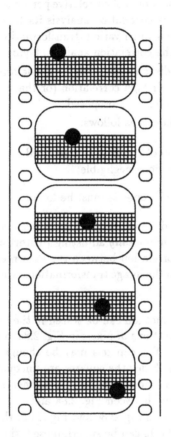

Figure 2.7: Flying ball in a cinema sequence

translations of A in Fig. 2.1 generate the contour of B, which is neither continuous, nor unidimensional. If instead of B we considered C, the outcome would be open (not closed).

2.4   Method of Variable Resolution

An implementation of the model of correlative perception requires considerable computing resources. The correlation analysis itself is rather slow and moreover it has to be performed under various transformations of data. Therefore, it is impossible to realize the correlation analysis, trying all possible distortions.

To end of realizing a directional search for the deformations of images (messages) which provides their high correlation (or that of certain submessages), a *method of variable resolution* is proposed.

The idea of the method is as follows:

- First, the resolution of the images is reduced in order to make the effect of small transformations negligible;

- Next, the correlated elements must be found;

- After the correlated elements have been discovered, the resolution is restored gradually while *locally* adjusting (distorting) the images in order to maintain the correlation between the known elements. These distortions correspond to the image transformations which provide the required high correlation.

The operation of the method can be traced in the following example. Consider two similar configurations of pixels in Fig. 2.8a–b coded by 1s on a background of dots (zeros). These pixels may be thought of as vertices of two squares of different sizes. Since the squares are unequal, the correlation function has no salient peaks. Indeed, under displacements and rotations, no pair of 1s can be superimposed between Fig. 2.8a and Fig. 2.8b. However, after a reduction in resolution, as in Fig. 2.8c and Fig. 2.8d, the two images are correlated. Indeed, some of the 1s can be superimposed; they are shown by frames. After the correlated elements have been discovered, the correlation can be improved by locally distorting neighborhoods of the correlated elements. Having determined the distortions which provide the highest correlation of the two images, one obtains the required deformation which transforms one original image into another, i.e. under which the two images in Fig. 2.8a–b are most correlated. If necessary, reducing and restoring the resolution can be realized gradually, in several steps.

One can see that the principal advantage of the method is that the search for *global distortions* is reduced to *local adjustments*. Note that reducing the

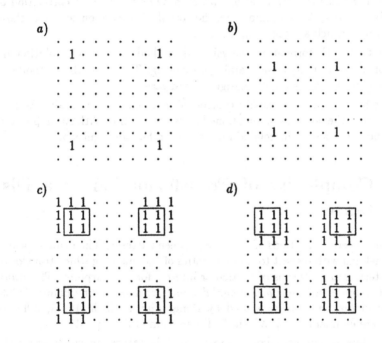

Figure 2.8: Illustration to the method of variable resolution
(dots denote zeros)

resolution (filtering) is usual in recognizing similarity; e.g. see (Palmer 1983; Witkin 1983; Bouman & Liu 1991). On the other hand, note the similarity of the method of variable resolution to the operational principle used in perceptrons (Minsky & Papert 1988) and pyramidal data structures (Hummel 1987).

However, we go further, providing the filtering and correlation schemes with a kind of a double feedback which guides the directional search for the image deformations required. It is realized by applying the correlation criterion which reveals the deformations which provide the highest correlation of successive images. At the same time, the complexity criterion sorts out those deformations which are too complex.

Note that both correlation analysis and method of variable resolution are realizable on neuron nets with parallel processing. Therefore, the method can be quite efficient on a specially designed hardware.

Since neuron nets are usually considered as models of the brain (Rossing 1990, p. 164), the observation mentioned is also compatible with our hypothesis about the existence of the related mechanism in human cognition.

2.5 Complexity of Transformation as a Distance

There are several definitions of distance between patterns. The distance which can be defined with respect to characteristics of the mapping which transforms one pattern into another is of particular interest for our purposes. The *translational distance* and *transformational distance* are discussed by Palmer (1983, p. 289); *distortion measure* is used by Rangarajan & Shah (1991); different metrics are enumerated by Witkin & Tonenbaum (1983b, p. 517).

According to our complexity approach, it is natural to characterize the distance between two patterns in terms of complexity of the deformation which transforms one pattern into another. Such a distance has the same properties as the metrical distance:

- **(Positivity)** The complexity of deformation is equal to 0 if and only if two patterns are identical, otherwise it is positive.

  Indeed, if two patterns are equal, the deformation required is trivial, implying its complexity to be equal to zero. If two patterns are dissimilar, the deformation required is significant, implying its complexity to be significant too.

- **(Symmetry)** The complexity of deformation (pointwise defined) equals to the complexity of the inverse deformation.

Indeed, if we consider the deformation of patterns as a pointwise correspondence, the complexity of the inverse transformation requires the same memory storage as the direct one.

- **(Triangle inequality)** The complexity of a superposition of two deformations (the sum of two complexities) is less than or equal to the complexity of the resulting direct deformation.

The complexity of deformation is well adapted to estimating the dissimilarity of patterns. For example, two patterns may be identical but remote in the Euclidean space. Since the complexity of deformation is small, their similarity is recognizable. Any distance based on absolute scales will fail in recognizing such an identity.

Moreover, learning can change the "distance" between patterns, providing some transformations with standard descriptions (e.g. rotations). Even if complex themselves, the reference complexity of these standard transformations can be small since their use requires calling related algorithms with a few parameters (cf. the call **R012** in Fig. 2.4).

In a sense, the complexity of deformation is complementary to the correlation coefficient, being a measure of dissimilarity instead of the measure of identity. However, using these two concepts is not equivalent. Indeed, two objects may be similar but not identical. In such a case the correlation analysis fails in recognizing similarity (because there may be no identity at all, as in Fig. 2.8a–b), whereas the complexity of the deformation, being small, indicates at the similarity (as recognized in Fig. 2.8d).

Note that the two complementary concepts, the correlation between two objects and the complexity of transformation of one into another, correspond to the two levels of our model of data representation: At the first level we have correlating generative elements, and the second level is reserved for the description of their deformations.

2.6 Distinctions of the Model

The need for an axiomatic theory of perception was realized while developing models of perceptual organization (Rock 1983, pp. 328–335; Palmer 1983; Leyton 1986).

In a sense, we develop an axiomatic approach (not a theory yet) to perception modeling. Postulating the self-organization capacity of perception aimed at data reduction, we attempt to derive its much less evident properties like: The capacity to segregate patterns, to arrange patterns into hierarchies, to recognize the causality in pattern generation, etc. While investigating hierarchical representations of data, we pose the following questions:

- Why a hierarchy?

- Which hierarchy? and

- How does the hierarchy correspond to the reality?

From the standpoint of our approach to minimizing the complexity of representations, the answers to these questions are, respectively:

- The hierarchization makes data representations compact.

- Consequently, a better hierarchy is the one which makes the data representation least complex.

- Under certain assumptions such a hierarchy reveals perception patterns and causal relationships in the data, making the first step towards their semantical description.

In our model, the self-organization capacity of perception is characterized in terms of optimal data representation but not in terms of recognition capabilities. We argue that the pattern recognition is determined by the criterion of least complexity, whereas most of recognition systems are based on threshold criteria which are adjusted at the learning phase.

Next, we elaborate the approach to separating patterns with respect to their *similarity*. Usually, the recognition of patterns is based on their classification with respect to their dissimilarity which is fixed by threshold criteria. Since we don't recognize dissimilarity, we also don't need thresholds for pattern separation.

Comparing the two approaches to pattern recognition, ours and the one based on threshold criteria, we see that the latter has two principal disadvantages:

- It requires a learning stage in order to determine thresholds,

- thresholds, being determined, make the recognition hardly adaptable to new circumstances (or additional learning is needed).

Unlike threshold criteria, the criterion of least complexity is self-adjustable to current circumstances. It guides the self-organization of data, not requiring any learning stage. This may be important in ambiguous cases where threshold criteria can fail. For instance, it is difficult to incorporate a threshold criterion in example illustrated by Fig. 2.4 in order to judge whether a sequence of durations should be interpreted as a single or repetitive pattern.

An important distinction of our approach is the way of recognizing the similarity. Usually, the measure of similarity is estimated by the degree of identity; this is the main idea of correlation analysis. On the contrary, we measure the

Table 2.2: Artificial intelligence and artificial perception in pattern recognition

	Artificial perception	Artificial intelligence
Function	Object segregation	Object identification
Performance	Data representation	Knowledge representation
Principle	Self-organization	Classification
Means	Data processing	Learning
Cues	Similarity	Dissimilarity
Criterion	Optimal representation	Threshold adjustment
Measure of similarity	Complexity of transformation	Correlation

difference between the objects by the degree of dissimilarity measured by the complexity of the deformation which is necessary to make objects identical. Such an approach is useful in cases when there is a similarity but there is no identity.

The enumerated features distinguish our artificial perception approach to pattern recognition from that of artificial intelligence which is traditionally based on learning, representation of knowledge, and classification of patterns. The complementarity of artificial intelligence and artificial perception in pattern recognition, as understoc 1 in our model, is shown in table 2.2.

2.7 Summary of the Chapter

Thus we enumerate the main distinctions of the proposed approach to perception modeling.

1. We consider a perception mechanism, the correlativity of perception, which is the interaction of two principles of data self-organization, "common fate" principle and simplicity principle. In our model these two grouping mechanisms control each other, resulting in a new quality of grouping.

2. The "common fate" principle is modeled in terms of generative elements and their transformations, corresponding to stable configurations of stimuli and relationships between them. Thus the structure is recognized with respect to replications but not with respect to dissimilarity.

3. The simplicity principle is formalized by the Kolmogorov criterion of optimal data representation (least memory storage required). By constructing optimal representations of data we attempt to segregate patterns and to reveal the causality in the data generation.

4. The search for stable configurations of stimuli is realized as finding similar messages in data arrays by means of correlation and autocorrelation analysis applied to the data arrays under their various distortions.

5. In order to realize the directional search for similar configurations of stimuli, the method of variable resolution is proposed. At first it reveals the similarity in general, and then localizes the search for fine matching.

6. The distance between two patterns is defined to be the measure of complexity of the transformation of one pattern into another. This measure can be modified while using some standard transformations (shifts, rotations, etc.) which are not to be described each time but simply coded with their parameters.

Chapter 3

Substantiating the Model

3.1 On the Adequacy of the Model

In the previous chapter the principle of correlativity of perception is formulated in its general form. As already said, by correlativity of perception we understand its capability to discover similar configurations of stimuli and to form high-level configurations from them. We justify such a representation of data, supposing that in most practical cases it is less complex (requires less memory) than the totality of data.

In particular, a chord spectrum can be regarded as generated by a tone spectral pattern translated along the $\log_2$-scaled frequency axis according to some interval structure. Therefore, the main idea of our approach to chord recognition is based on finding similar groups of sinusoidal tones in the chord spectrum with which the chord can be described as an acoustical contour drawn by some generative subspectrum. We treat audio data by analogy with the visual data in Fig. 2.1 and 2.2. This analogy is displayed in Table 3.1. Note that even if the pitch is not identifiable (which corresponds to the impossibility to recognize tones like the unknown symbol II in Fig. 2.2), the chord can be recognized by relationships of the patterns of lower level (corresponding to the recognizability of symbol B in Fig. 2.2).

Table 3.1: Correspondence between visual and audio data

	In statics		In dynamics	
	Visual data	Audio data	Visual data	Audio data
Stimuli	Pixels	Partials	Pixels	Partials
Low-level patterns	Symbols A	Notes	Object	Notes
High-level pattern	Symbol B	Chord	Trajectory	Melody

Obviously, storing a tone spectrum, say, with 10 partials, and two intervals of a major triad is more efficient than storing about 30 partials of the chord spectrum. However, the efficiency of such a representation is only a conjecture. It may happen that a chord spectrum is representable in terms of generative tone spectra in several ways, implying the ambiguity in the chord recognition, or the optimal (least complex) representation may differ from the chord decomposition into notes.

Therefore, in order to substantiate our model, we must be sure that:

- the description of a chord spectrum as generated by translations of a tone spectrum corresponds to the chord note structure as it is perceived;

- such a description provides the optimal representation of spectral data;

- this optimal representation is unique.

In the present chapter we prove the above items for two-tone intervals, major triads, and minor triads. This way we show that, indeed, the representation of a chord spectrum in terms of generative spectral patterns and their translations is unique, optimal, and corresponds to the note structure of the chord. In other words, we prove that our recognition model is adequate to human perception.

In Section 3.2, "Tone Spectra", we recall some basic facts from musical acoustics. We give definitions of periodical waveforms, partial tones, their amplitude and phase, and tone spectra, both discrete and continuous. Discrete spectra are considered in terms of Dirac delta-functions. Finally, the distinction between harmonic and inharmonic tones is explained.

In Section 3.3, "Representation of Tones", we enumerate the conditions for tone spectra which are assumeed in our study. We define a discrete audio power spectrum which corresponds to the one perceived by humans. For this purpose the frequency axis is logarithmically scaled and divided into a finite number of frequency bands, the phase of partial tones is ignored, and their amplitude measured by positive integers. These definitions are illustrated with examples which are used in the sequel.

In Section 3.4, "Generation of Chord Spectra", we recall the notion of convolution. We show that a chord spectrum which is generated by multiple translations of a tone spectral pattern along the $\log_2$-scaled frequency axis can be written down as a convolution product of two spectra. The first factor is a tone spectrum which is associated with a note. The second factor which is associated with the translations of the tone spectral pattern is said to be the interval distribution of the chord.

In Section 3.5, "Unique Deconvolution of Chord Spectra", we establish an isomorphism between polynomials over non-negative integers and discrete audio spectra with respect to addition and convolution. Unlike polynomials over integers, polynomials over non-negative integers have no unique factorization property, and unique deconvolution of discrete audio spectra doesn't take place in the general case. However, for spectra of two-tone intervals and major or minor triads generated by harmonic tones (with pitch salience) the unique deconvolution is valid. Hence, we obtain that the only deconvolution of a chord spectrum equals to its generation.

In Section 3.6, "Causality and Optimal Data Representation", we prove that the deconvolution of a chord spectrum into a tone spectrum and an interval distribution provides the least complex representation of spectral data. Since such a deconvolution is unique, the optimal representation of spectral data reveals the causality in their generation. From this we conclude that, in case of chord recognition, the optimal description contributes to the recognition of the causality in sound.

In Section 3.7, "Interpretation of the Results", we show that the $\log_2$-scaling of the pitch axis as well as the insensitivity of the ear to the phase of the signal are essential for adequate perception of physical reality. In particular, these properties provide for the conditions for the perception of musical tones as entire sound objects rather than as compounds of sinusoidal tones (in our model this corresponds to irreducibility of tone spectra). Besides, these assumptions are important for the capacity to recognize chords as compounds of musical tones. In other words, these conditions are necessary for recognizing the structure in audio scenes, corresponding to the physical causality in the acoustical environment.

In Section 3.8, "Main Items of the Chapter", we summarize the contents of the chapter.

3.2 Tone Spectra

In this section we recall some facts about tone representation. For more details one can refer to systematical introductions to musical acoustics, see e.g. Helmholtz (1877), Benade (1976), Roederer (1975), and books on signal processing, see e.g. Oppenheim & Schafer (1975); Rabiner & Gold (1975). A brief tutorial of related mathematical formulas is given by Moore (1978).

By a *musical tone*, or *harmonic tone* we understand a sound with a clear pitch salience. Such a sound is generated by a periodical, or quasi-periodical oscillation of the air pressure. Its periodical frequency, said to be the tone's *fundamental frequency* is usually associated with the tone's *pitch*.

It is known that the voices of musical instruments have a complex structure. For example, consider a cello string. Vibrating by its whole length, it produces a sinusoidal tone with fundamental frequency ω. Yet the string vibrates by its halfs, thirds, etc., producing sinusoidal tones with frequencies $2\omega, 3\omega$, etc., respectively, which are said to be *partial tones* or simply *partials*, or *overtones*.

The pitches of the first 16 partials from the *overtone series* of tone C (*do* of grande octave) are shown in Fig. 3.1. All these partials are summarized with different intensities, resulting in a complex oscillation perceived as the cello sound.

Note that the multiplication of the fundamental frequency implies the same difference between all the frequencies of successive partials of musical tones. This difference which is equal to the fundamental frequency ω. However, the graphical distance at the stuff in Fig. 3.1 between the pitches of successive partials is getting less while increasing in their number. For example, the interval between the first two partials is an octave, whereas the interval between the second and the third partials is a fifth, etc. This is the implication of the logarithmic scale of pitch in musical notation. If a linear scale of pitch was used, then every next octave would be twice larger than the preceding one.

In signal processing the frequency axis is commonly linearly scaled. However, $\log_2$-scaled frequency axis is better suited for our purposes. Also note that such a scale is also inherent in audio perception (Gelfand 1981). It implies that equal distances on the scale correspond to equal musical intervals. The signal processing aspects of logarithmic pitch scaling are discussed by Teaney, Mourizzi, & Mintzer (1980).

According to Fourier's theorem, any tone with a periodic waveform is a sum of *harmonics*, i.e. partials with the frequencies ω_k which satisfy the *harmonic* frequency ratio

$$\omega_1 : \omega_2 : \ldots : \omega_k : \ldots = 1 : 2 : \ldots : k : \ldots. \tag{3.1}$$

Hence, any periodic waveform $f(t)$ which is a function of time t can be represented as follows

$$f(t) = \sum_{k=1}^{\infty} a_k \sin(2\pi\omega_k t - \varphi_k), \tag{3.2}$$

where

$\omega_1 = \omega$ is the fundamental frequency of the waveform's oscillation which is usually associated with the tone's *pitch*;

$\omega_k = k\omega$ is the kth partial's frequency which is k times the fundamental frequency ω;

a_k is the *amplitude* of the kth partial which is interpreted as the *loudness* of the given partial;

φ_k is the *phase* of the kth partial which is interpreted as the "entry delay" of the given partial.

The formula (3.2) establishes a one-to-one correspondence between the waveform $f(t)$ and the set of characteristics of the associated sinusoidal overtones. Since the range of perceptible frequencies is approximately limited to

Figure 3.1: The overtone series

the band 20–$20000\,Hz$, the infinite series (3.2) can be always reduced to a finite sum. This means that *all* perceptible information about a musical tone is represented by a finite number of parameters.

To be more precise, the one-to-one correspondence mentioned is established between periodical functions and their spectra. The *spectrum* of a function $f(t)$ is defined to be the complex-value function

$$S(\omega) = \int_{-\infty}^{+\infty} f(t)e^{-i\omega t}\, dt.$$

If $f(t)$ is a signal, its spectrum can be understood as the density function of different sinusoidal constituents of the signal $f(t)$. Since spectral values are complex numbers, they are interpreted in terms of amplitude and phase of the related sinusoidal constituents of the signal $f(t)$.

If a function $f(t)$ satisfies some rather general conditions, then it is uniquely reconstructed from its spectrum as follows

$$f(t) = \frac{1}{2\pi} \int_{-\infty}^{+\infty} S(\omega)e^{i\omega t}\, d\omega. \tag{3.3}$$

This formula justifies the interpretation of the spectrum as a density of sinusoidal constituents of the signal. In fact, by virtue of (3.3), the signal $f(t)$ is a "weighted sum" of sinusoidal partials with frequencies ω with the complex "weight coefficients" (meaning amplitude and phase) $S(\omega)$.

In case of a finite number of frequencies $\omega_1, \ldots, \omega_K$ which are present in the signal, the spectral density is concentrated at these frequencies. Then the spectrum is *pointwise*, being a sum of impulses

$$S(\omega) = \sum_{k=1}^{K} a_k \delta(\omega - \omega_k), \tag{3.4}$$

where the impulses are given by the Dirac *delta function* $\delta(x)$ conventionally defined to be

$$\delta(x - a) = \delta_a = \begin{cases} +\infty & \text{if } x = a, \\ 0 & \text{if } x \neq a, \end{cases} \tag{3.5}$$

so that

$$\int_{-\infty}^{+\infty} \delta(x - a)\, dx = 1. \tag{3.6}$$

Substituting spectrum (3.4) into (3.3), we obtain the representation (3.2). Moreover, the signal is periodical if and only if the frequency ratio of its partials is harmonic, i.e. satisfies the condition (3.1).

Usually, pointwise spectra are displayed by a series of finite (!) impulses whose height is proportional to the amplitude of related partial tones versus the frequency axis. Two such spectra are shown in Fig. 3.2b–c where the frequency axis is logarithmically scaled.

Besides harmonic tones characterized by the frequency ratio (3.1), we consider sounds with no pitch salience. These sounds can be of two types:

- *inharmonic tones*, e.g. bell-like sounds,

- *noises*, e.g. drum sounds.

Both harmonic and inharmonic *tones* are characterized by pointwise spectra. Therefore, inharmonic tones can be represented by the series (3.2) with the only difference that the ratio of partial frequencies ω_k is *inharmonic* (i.e. not harmonic).

An example of an inharmonic ratio is

$$\omega_1 : \omega_2 = 1 : \sqrt{2}.$$

Since the above frequency ratio cannot be expressed in integers, the associated partials cannot be considered as some two partials of a harmonic tone. In particular, this implies that the signal composed by these two partials is not periodical and therefore has no salient pitch.

A noise is characterized by a *continuous* spectrum which spreads over a certain frequency band. Since all frequencies from a certain band are present, no countable set of partials is sufficient to describe the signal. Therefore, the integral representation (3.3) is needed instead of the series (3.2).

However, in discrete computational models both inharmonic tones and noises are approximated by finite sums of overtones. Therefore, in computing the distinction between inharmonic tones and noises is conditional. Recall that in discrete representations the frequency axis is divided into bands where the signal characteristics are measured. If a spectrum has constituents from successive frequency bands, such a spectrum can be considered as continuous. If a spectrum has no constituents from any pair of successive frequency bands, such a spectrum can be considered as discrete. Obviously, such a judgement depends greatly on the accuracy of discretization. Indeed, two close partials can be unseparable under a poor accuracy while being separable under a refined accuracy, implying the spectrum to be "continuous" in the former case and "discrete" in the latter.

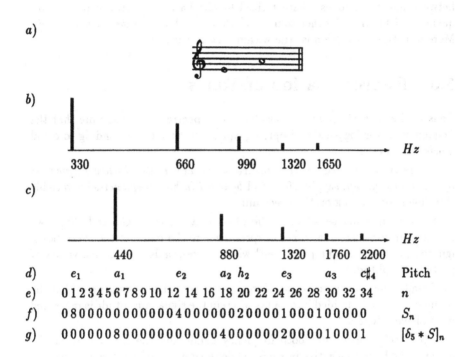

Figure 3.2: Example of tone spectra

a) tones e_1 and a_1 in standard musical notation;

b) the audio spectrum (with $\log_2$-scaled frequency axis) of tone e_1 for a harmonic voice with 5 successive partials which have decreasing power;

c) the same for tone a_1;

d) the pitches $e_1, \ldots, c\sharp_4$ which correspond to the mean frequencies of the frequency bands for the given spectral representations;

e) the indexes $n = 1, \ldots, 34$ of frequency bands for the given discrete spectra;

f) the discrete audio spectrum of tone e_1 under the frequency resolution within a semitone (1/12 octave) — string S_n;

g) the same for the second tone — string $[\delta_5 * S]_n$.

Since we deal with discrete models, we make the only principal distinction between harmonic tones, characterized by the harmonic ratio of partial frequencies (3.1), and all other sounds which are said to be *inharmonic sounds*. Moreover, this distinction is true within certain accuracy.

3.3   Representation of Tones

Thus in the present study we consider *audio* spectra, i.e. we assume that the frequency axis is $\log_2$-scaled, implying equal distances corresponding to equal musical intervals.

We restrict ourselves to the spectra which are *limited in low frequencies* by a fixed threshold, say, by $20\,Hz$ and *bounded in high frequencies* by variable thresholds, depending on the spectrum.

Since man is not sensitive to the phase of the signal (Gelfand 1981), we restrict our consideration to *power spectra* where the phase of the overtones is ignored (and well as the phase 180^0 which corresponds to negative values of partials' amplitude).

Finally, the spectra to be considered are *discrete*, i.e. it is assumed that the frequency range is divided into bands wherein the signal amplitude is measured by positive integers.

All of this means that both frequency bands and spectral values can be enumerated by non-negative integers, while the number of bands with a strictly positive level being always finite.

Thus by *spectra* we understand the expressions of the form

$$S = S(x) = \sum_{n=0}^{N} S_n \delta(x - n) = \sum_{n=0}^{N} S_n \delta_n, \qquad (3.7)$$

where

N is the total number of frequency bands minus 1;

n is the index of a frequency band;

δ_n is the Dirac delta function defined in (3.5–3.6), i.e. the unit impulse at the nth frequency;

S_n is a non-negative integer interpreted as the signal power in the nth frequency band.

The *support* of spectrum (3.7) is defined to be the set of its partial frequencies. To be more precise, it is defined as the set of indexes of frequency bands as follows

$$\Delta_S = \{n : S_n \neq 0\}.$$

Since by assumptions we consider bounded spectra, the spectral supports are always finite.

By virtue of (3.1), the kth partial of a harmonic tone falls into the frequency band with the following index

$$n_k = p + [C \log_2 k + 0.5], \qquad (3.8)$$

where

$p = n_1$ is the index of the frequency band with the tone's *fundamental frequency*;

C is the constant which characterizes the accuracy of spectral representation, being equal to the number of frequency bands per octave,

$[\cdot + 0.5]$ is the *rounding function* (since $[\cdot]$ retains the integer part of its argument, function $[\cdot + 0.5]$ retains the closest integer to a given real number).

A spectrum with the partials which satisfy the condition (3.8) is said to be *harmonic*. It is easy to see that a discrete audio spectrum S is harmonic if and only if its support Δ_S has only harmonic frequencies, i.e. for a certain K it holds

$$\Delta_S \subset \{n_k : n_k = [C \log_2 k + 0.5], \ k = 1, \ldots, K\}. \qquad (3.9)$$

Two harmonic spectra are shown in Fig. 3.2.

3.4 Generation of Chord Spectra

First of all, recall that the *convolution* of two functions $f(t)$ and $g(x)$ is defined to be the function

$$[f * g](t) = \int_{-\infty}^{+\infty} f(t - x) g(x) \, dx.$$

Under rather general conditions the convolution is commutative, associative, and linear with respect to both arguments f and g.

As follows from the definition of delta function (3.5–3.6),

$$[f * \delta_a](t) = \int_{-\infty}^{+\infty} f(t - x) \delta(x - a) \, dx = f(t - a),$$

corresponding to the translation of function f by a to the right along the t-axis.

Consequently, the delta function $\delta = \delta_0$ is a *convolution unit*, since for all functions f it holds

$$\delta * f = f * \delta = f.$$

Also note that for arbitrary a and b we have

$$\delta_a * \delta_b = \delta_{a+b}.$$

Since we consider a $\log_2$-*scaled* frequency axis, a pitch shift of a complex tone spectrum corresponds to a *parallel* translation of the whole tone spectrum along the frequency axis. A translation of a spectrum (3.7) by m bands to the right is given by the convolution

$$\delta_m * S = \sum_n S_n \delta_{n+m} \qquad (3.10)$$

An example of a translation of a harmonic spectrum is shown in Fig. 3.2.

As fixed by Conjecture 4 from Section 1.4, a chord spectrum can be regarded as generated by a multiple translation of a tone spectrum. By virtue of (3.10), a chord spectrum S can be written down as follows

$$S = \sum_m [I_m \delta_m * T] = [\sum_m I_m \delta_m] * T = I * T, \qquad (3.11)$$

where T is a tone spectrum, and

$$I = \sum_m I_m \delta_m$$

is said to be the *interval distribution* of the chord.

For example, the interval distribution associated with the interval of m semitones is defined to be

$$I_m = \delta_0 + \delta_{[mC/12+0.5]}.$$

If the accuracy of frequency resolution is the same as the accuracy of standard note stuff, i.e. to within one semitone ($C = 12$), then the major triad (4 semitones) has the interval distribution

$$I_4 = \delta_0 + \delta_4. \qquad (3.12)$$

Now consider a chord which is formed by two intervals of m and k semitones from the lowest tone. The corresponding interval distribution is as follows

$$I_{m,k} = \delta_0 + \delta_{[mC/12+0.5]} + \delta_{[kC/12+0.5]}.$$

If the accuracy of frequency resolution is to within a quarter of a tone ($C = 24$), then the interval distribution of the major triad which is built by major third and fifth (4 and 7 semitones) has the form

$$I_{4,7} = \delta_0 + \delta_8 + \delta_{14}.$$

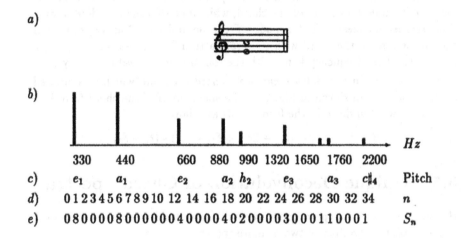

a)

b)

| 330 | 440 | 660 | 880 | 990 | 1320 | 1650 | 1760 | 2200 | *Hz* |

c) e_1 a_1 e_2 a_2 h_2 e_3 a_3 $c\sharp_4$ Pitch

d) 0 1 2 3 4 5 6 7 8 9 10 12 14 16 18 20 22 24 26 28 30 32 34 n

e) 0 8 0 0 0 0 8 0 0 0 0 0 4 0 0 0 4 0 2 0 0 0 3 0 0 0 1 1 0 0 0 1 S_n

Figure 3.3: Example of chord spectrum

a) chord (e_1, a_1) in standard musical notation;

b) the audio spectrum (with $\log_2$-scaled frequency axis) of the chord for a harmonic voices with 5 successive partials which have decreasing power;

c) the pitches $e_1, \ldots, c\sharp_4$ which correspond to the mean frequencies of the frequency bands for the given spectral representations;

d) the indexes $n = 1, \ldots, 34$ of frequency bands for the given discrete spectra;

e) the discrete audio spectrum of the chord under the frequency resolution within a semitone (1/12 octave) — string S_n.

A spectrum S is said to be *simple* if all its coefficients are relatively prime (in the sense of number theory) and its first coefficient $S_0 \neq 0$.

Obviously, the decomposition (3.11) can be written down as follows

$$S = a\delta_p * T * I, \tag{3.13}$$

where the interval distribution I and tone spectrum T are simple.

A simple interval distribution I corresponds to the intervals between the lowest note and other notes of the chord, while its coefficients I_m determining their *relative loudness*. The term $a\delta_p * T$ can be understood as a spectrum of the lowest tone of the chord with spectral pattern T, loudness a, and pitch p (the index of the frequency band with the tone's fundamental frequency).

For example, in Fig. 3.3 one can see a chord spectrum which is composed of two tone spectra shown in Fig. 3.2. The spectrum S of the chord from Fig. 3.3 can be written down in the form (3.13) as follows

$$S = \delta_1 * (8\delta_0 + 4\delta_{12} + 2\delta_{19} + \delta_{24} + \delta_{28}) * (\delta_0 + \delta_5).$$

3.5   Unique Deconvolution of Chord Spectra

By *deconvolution* of a spectrum S we shall understand a representation of S as a convolution product of two or more spectra

$$S = S_1 * \ldots * S_k,$$

where no factor S_i is a convolution unit, i.e. δ multiplied by a constant a.

In order to investigate the items enumerated in section 3.1, we pose the question: Given a chord spectrum S generated according to (3.13); does there exist a deconvolution of S into simple spectra other than (3.13)?

Lemma 1 (Isomorphism Between Discrete Spectra and Polynomials over Integers) *Define the correspondence between discrete spectra with integral coefficients and polynomials over integers by the equality of their coefficients*

$$S = \sum_{n=0}^{N} a_n \delta_n \quad \longleftrightarrow \quad p(x) = \sum_{n=0}^{N} a_n x^n.$$

Then this correspondence is one-to-one, the sum of two spectra corresponds to the sum of the associated polynomials, and the convolution of two spectra corresponds to the product of the associated polynomials.

PROOF OF LEMMA 1. Obviously, the correspondence is one-to-one. Consider two spectra and the associated polynomials

$$S = \sum_{n=0}^{N} a_n \delta_n \quad \longleftrightarrow \quad p(x) = \sum_{n=0}^{N} a_n x^n$$

$$T = \sum_{n=0}^{M} b_n \delta_n \longleftrightarrow q(x) = \sum_{n=0}^{M} b_n x^n.$$

Then

$$S + T = \sum_{n=0}^{\max\{M,N\}} (a_n + b_n)\delta_n \longleftrightarrow \sum_{n=0}^{\max\{M,N\}} (a_n + b_n)x^n = p(x) + q(x).$$

Since

$$x^{i+j} = x^i x^j \longleftrightarrow \delta_i * \delta_j = \delta_{i+j},$$

all the difference between multiplication of polynomials and convolution of spectra is in manipulating lower indexes of δ instead of powers of x. Hence, we have

$$S * T = \sum_{n=0}^{M+N} \left(\sum_{i+j=n} a_i b_j \right) \delta_n \longleftrightarrow \sum_{n=0}^{M+N} \left(\sum_{i+j=n} a_i b_j \right) x^n = p(x)\, q(x),$$

as required. ∎

By analogy with polynomials, a spectrum S is said to be *irreducible* if it cannot be factored into a convolution product of two spectra, each other than convolution unit δ multiplied by a constant a.

Referring to the unique factorization of polynomials over integers (Birkhoff & Mac Lane 1965), we obtain the unique deconvolution of spectra with integral coefficients. Since we are interested in the deconvolution of discrete spectra with *non-negative* integral coefficients into spectra with *non-negative* integral coefficients, the question of unique deconvolution is reformulated for polynomials with the same property.

Ambiguous factorization into primes arises in special number fields, see Pollard (1950), p. 76, and, as shown at the end of this section, ambiguous factorization is also inherent in polynomials over non-negative integers. However, the polynomials associated with discrete power spectra of usual chords, like two-tone intervals and major or minor triads, are uniquely factored into irreducible polynomials. In order to prove the unique deconvolution theorem for chord spectra, we need several propositions.

Lemma 2 (Sufficient Condition for Irreducible Spectra) *Consider a simple spectrum S whose support Δ_S contains two points at least, i.e.*

$$S = \sum_{k=1}^{K} a_k \delta_{\omega_k}, \quad \omega_1 < \ldots < \omega_K, \ K \geq 2.$$

Suppose that the distance between the last two partials (first two partials) of S is less than the distance between any other pair of partials of S, i.e.

$$d = \omega_K - \omega_{K-1} < \omega_k - \omega_{k-1}, \quad k = 2, \ldots, K-1 \tag{3.14}$$

$$(d = \omega_2 - \omega_1 < \omega_k - \omega_{k-1}, \quad k = 3, \ldots, K). \tag{3.15}$$

Then S is irreducible.

PROOF OF LEMMA 2. Firstly, assume that the distance d between the last two partials in S is less then the distance between any other pair of partials. Since S is simple, the factorization $S = \delta_p * S'$, where $p > 0$, is impossible. Consequently, if $S = I * T$ then each of the supports Δ_I and Δ_T should contain two points at least. Since the partials indexed by ω_{K-1} and ω_K are the last in the spectrum S, and operator $I * (\cdot)$ performs parallel translations of spectrum T, only the following two cases are possible.

1. The partials indexed by ω_{K-1} and ω_K result from the two last partials of spectrum T translated. Then the distance between the last two partials of T is also equal to d. Since by assumption Δ_I contains more than one point, the spectrum T is repeated in S several times. Consequently the same distance between partials d appears elsewhere in S, which contradicts the condition (3.14).

2. The partials indexed by ω_{K-1} and ω_K result from two translations of the last partial of spectrum T. Then the distance between the last two points in Δ_I is equal to $\omega_K - \omega_{K-1}$. Since by assumption Δ_T contains more than one point, the same distance between points should appear elsewhere in Δ_S, resulting from the same translation of the first partial of T. This however contradicts the condition (3.14).

Thus we have shown the impossibility of convolution factoring S into two non-trivial spectra. One can easily modify the proof for the assumption (3.15). ■

Lemma 3 (Irreducibility of Harmonic Spectra) *Let T be a simple harmonic spectrum or a simple segment of a harmonic spectrum, i.e.*

$$T = \sum_{k=K_1}^{K} a_k \delta_{\log_2 k}, \quad 1 \leq K_1 < K.$$

Then T is irreducible.

PROOF OF LEMMA 3. Consider the support of spectrum T. Note that the difference $\log_2 k - \log_2(k-1)$ decreases monotonically with an increase in k. Consequently, the distance between the last two partials of T is less than the distance between any other pair of partials. Then by Lemma 2 we obtain the required statement. ■

Note that in case of discrete spectra the above lemma is valid, if the spectral resolution is sufficiently accurate to distinguish that the distance between the last two partials is less than the distance between any other pair of partials.

Lemma 4 (Irreducibility of Intervals and Triads) *Simple interval distributions, corresponding to two-tone intervals, major triads, or minor triads, are irreducible.*

PROOF OF LEMMA 4. The required fact for intervals follows from Lemma 2. Since the major and minor triads are formed by major and minor thirds, the corresponding interval distributions satisfy the assumption of Lemma 2. ∎

Obviously, the above lemma is valid for discrete spectra as well.

Also note that the assumption of non-negativity of spectral coefficients is important. Otherwise, as seen from the following example, even the interval of major triad is not irreducible.

Example 1 (Reducibility of Major Triad Interval Distribution) Let the spectral accuracy be within one semitone ($C = 12$). Hence, the major triad corresponds to 4 frequency bands. Then by virtue of (3.12) and Lemma 1, the reducibility of the major triad follows from the polynomial factorization

$$x^4 + 4 = (x^2 + 2x + 2)(x^2 - 2x + 2). \tag{3.16}$$

Further we shall divide a spectrum

$$S = \sum_{n=1}^{N} a_n \delta_{\omega_i},$$

whose partial (impulses) tones have frequencies

$$\omega_1 < \omega_2 < \ldots < \omega_N,$$

into lower and higher parts, which are said to be head and tail, as follows:

$$S = \sum_{n=1}^{N} a_n \delta_{\omega_n} = \sum_{n=1}^{N-Q} a_n \delta_{\omega_n} + \sum_{n=N-Q+1}^{N} a_n \delta_{\omega_n},$$

Under these conventions the spectrum's higher part with Q partials

$$S_Q = \sum_{n=N-Q+1}^{N} a_n \delta_{\omega_n}$$

is said to be the *Q-tail of S*.

We say that two spectra S and T have *congruent Q-tails* if

$$S_Q = \delta_a * T_Q$$

for certain a. This will be denoted

$$S_Q \sim T_Q.$$

Lemma 5 (Unique Factorization for Spectral Tails) *Let T, I, U, and J be four spectra, each having more than one impulse, such that*

$$T * I = U * J.$$

Let d be the distance between the last two impulses of T and f be the distance between the last two impulses of I. Suppose that

$$d < f.$$

Define the tail T_Q by the given bandwidth f, i.e. determine Q so that

$$T_Q = \sum_{n=N-Q+1}^{N} a_n \delta_{\omega_n} = \sum_{\omega_n : \omega_n > \omega_N - f} a_n \delta_{\omega_n}, \qquad (3.17)$$

and denote the distance between the last and the Qth to last partial of T by

$$g = \omega_N - \omega_{N-Q} < f.$$

Then either $U_Q \sim T_Q$ and the distance between the last two partials of J is greater than or equal to g, or $J_Q \sim T_Q$ and the distance between the last two partials of U is greater than or equal to g.

PROOF OF LEMMA 5. Consider spectrum

$$S = T * I.$$

Since S is generated by translations of T at intervals from I, and the distance f between the last two impulses of I determines the size of T_Q, we have

$$S_Q \sim T_Q. \qquad (3.18)$$

In particular, the distance between the last two impulses of S_Q is equal to d which is the smallest distance between impulses of T_Q.

Consider the last two impulses of S. Since by assumption $f > d$, we have

$$S_2 \sim T_2.$$

By assumption U and J have at least two impulses each. Let us analyze several possibilities.

The case when the distance between the last two impulses of U is strictly smaller than d is impossible, since then the distance between the last two impulses of $S = U * J$ would be less than d.

By the same reasons, the distance between the last two impulses of J cannot be strictly smaller than d.

The case when both the distance between the last two impulses of U and the distance between the last two impulses of J are strictly greater than d is also impossible, since then the distance between the last two impulses of $S = U * J$ would be strictly greater than d.

Therefore, the only possible cases are the following:

(a) The distance between the last two impulses of U is equal to d and the distance between the last two impulses of J is strictly greater than d.

(b) The distance between the last two impulses of J is equal to d and the distance between the last two impulses of U is strictly greater than d.

Since the above two cases differ in renaming U and J, consider the first case. Let the partials (impulses) of S be indexed from the end of S by numbers $q = 1, \ldots, Q$ (the partial indexed by q is the qth to last partial of S). Now we shall show that every qth to last impulse of S results from the convolution of the qth to last impulse of U by the last impulse of J. We shall prove it by induction on $q = 2, \ldots, Q$.

For $q = 2$ the required statement is already proved. Suppose that it is valid for a certain positive integer $q < Q$. Obviously, for the $q + 1$st to last impulse of S there are only four possibilities:

1. The $q + 1$st to last impulse of S results from the convolution of the $q + 1$st to last impulse of U by the last impulse of J.

2. The $q + 1$st to last impulse of S results from the convolution of the $q + 1$st to last impulse of U by a not-last impulse of J.

3. The $q + 1$st to last impulse of S results from the convolution of the last impulse of J by an impulse of U whose number from the end is $r < q + 1$.

4. The $q + 1$st to last impulse of S results from the convolution of a not-last impulse of J by an impulse of U whose number from the end is $r < q + 1$.

Consider the second case. If the $q + 1$st to last impulse of S results from the convolution of the $q + 1$st impulse of U by a not-last impulse of J, then, since $S = U * J$, the distance d must be repeated in S_Q, which contradicts to the fact that by virtue of (3.18) the distance d is the smallest distance between partials of S_Q. Consequently, this case is impossible.

Consider the third case. By the statement of induction the convolution of the rth to last impulse of U by the last impulse of J is equal to the rth to last impulse of S. Consequently, the third case is impossible.

The fourth case is impossible by the same reasons as the second case.

Therefore, the only possible case is the first, whence $U_{q+1} \sim S_{q+1}$.

Thus we have shown that $U_Q \sim S_Q$, whence by virtue of (3.18) we have

$$U_Q \sim T_Q.$$

Since every impulse of S_Q results from the convolution of the last impulse of J by the corresponding impulse of U, the distance between the last two impulses of J is greater than or equal to g (otherwise the shortest distance d would be repeated in $S_Q \sim T_Q$). ∎

Lemma 6 (Uniqueness of Interval Decomposition) *Consider an interval distribution I with two impulses, and let the distance f between its impulses be smaller than or equal to 12 semitones, i.e.*

$$I = a_1 \delta_{\omega_1} + a_2 \delta_{\omega_2} \quad (\omega_1 < \omega_2);$$
$$f = \omega_2 - \omega_1 \leq 12 \text{ semitones.}$$

Let T be a harmonic spectrum with N successive partials, where N is odd and sufficiently large so that the distance between the last three harmonics of T is less than f, i.e.

$$T = \sum_{n=1}^{N} b_n \delta_{\omega_n} \quad (\omega_1 < \ldots < \omega_N);$$
$$f > \omega_N - \omega_{N-2}.$$

Consider spectrum $S = T * I$. Then its deconvolution into non-trivial factors (i.e. having at least two impulses each) is unique to within order and units.

PROOF OF LEMMA 6. Consider a deconvolution of S into two non-trivial factors U and J, i.e. $U * J = S$. By virtue of Lemma 5 we can assume that

$$U_3 \sim T_3.$$

We are going to show that $J = I$ whence it will follow that $U = T$.

At first let us show that J has two impulses as I has, and that the distance between these impulses is also f. Suppose that it is not true. Since by assumption J has two impulses at least, there exists an impulse in J whose distance from the highest impulse of J differs from f, and therefore the tail T_3 is repeated in S at a distance from the end of S different from f. We shall show that it is not possible.

Consider spectrum S as a union of two sets of harmonics. Let the first set correspond to the lowest tone with the fundamental frequency p. Then the frequencies of its harmonics are np, where $n = 1, \ldots, N$. Let the second set correspond to the upper tone with the fundamental frequency q. Then the frequencies of its harmonics are kq, where $n = 1, \ldots, N$. Since by assumption the interval considered is not greater than the octave, we have

$$p < q \leq 2p. \tag{3.19}$$

Since we suppose that T_3 is repeated in S at an interval different from f, there exists a positive frequency

$$r \neq p, q \tag{3.20}$$

such that the frequencies Nr, $(N-1)r$, and $(N-2)r$ are equal to some frequencies of spectrum S. Since Nq is the highest frequency of spectrum S, we have $Nr \leq Nq$, whence by virtue of (3.20)

$$r < q. \qquad (3.21)$$

Since we suppose that frequency Nr is inherent in spectrum S, there are two possibilities:

1. Nr is a harmonic frequency of the higher tone, i.e.

$$Nr = kq, \ k < N, \qquad (3.22)$$

(if $k = N$ we would have $r = q$ against (3.20)).

2. Nr is a harmonic frequency of the lower tone, i.e.

$$Nr = mp, \ m < N, \qquad (3.23)$$

(if $m = N$ we would have $r = p$ against (3.20)).

Let us analyze both possibilities.

1. Consider possibility (3.22). Since both frequencies $(N-1)r$ and $(N-2)r$ are inherent in spectrum S, we have the following cases.

   (a) $(N-1)r = k_1q$, where obviously $k_1 < k$.
       By virtue of (3.22) this gives

$$r = (k - k_1)q, \ k - k_1 \geq 1,$$

   whence $r \geq q$ against (3.21). Therefore, this case is impossible.

   (b) Consequently, it must be $(N-1)r = m_1p$ for a certain m_1.

   (c) $(N-2)r = m_2p$, where obviously $m_2 < m_1$.
       By virtue of (b), this gives

$$r = (m_1 - m_2)p, \ m_1 - m_2 \geq 1,$$

   whence either $r = p$ against (3.20), or $r \geq 2p$ against (3.19) and (3.21).

   (d) $(N-2)r = k_2q$, where obviously $k_2 < k$.
       By virtue of (3.22) this gives

$$r = \frac{k - k_2}{2}q, \ k - k_2 \geq 1.$$

The case $\frac{k-k_2}{2} \geq 1$ contradicts to (3.21), consequently $\frac{k-k_2}{2} = \frac{1}{2}$, which is equivalent to $k_2 = k - 1$. Substituting it into (c) and using (3.22), we obtain

$$\frac{N}{N-2} = \frac{k}{k-1}, \quad k - k_2 \geq 1,$$

which is equivalent to

$$1 + \frac{2}{N-2} = 1 + \frac{1}{k-1},$$

whence

$$N = 2k.$$

Since by assumption N is odd, this is impossible.

Therefore, the possibility $Nr = kq$ is excluded.

2. Consider possibility (3.23). Since both frequencies $(N-1)r$ and $(N-2)r$ are inherent in spectrum S, we have the following cases.

   (e) $(N-1)r = m_1 p$, where obviously $m_1 < m$.
   By virtue of (3.23) this gives

   $$r = (m - m_1)p, \quad m - m_1 \geq 1,$$

   whence either $r = p$ against (3.20), or $r \geq 2p$ against (3.19) and (3.21). Therefore, this case is impossible.

   (f) Consequently, it must be $(N-1)r = k_1 q$ for a certain k_1.

   (g) $(N-2)r = k_2 q$, where obviously $k_2 < k_1$.
   By virtue of (f) this gives

   $$r = (k_1 - k_2)q, \quad k_1 - k_2 \geq 1,$$

   whence $r \geq q$ against (3.21).

   (h) $(N-2)r = m_2 p$, where obviously $m_2 < m$.
   From this and (3.23) we obtain

   $$r = \frac{m - m_2}{2}p, \quad m - m_2 \geq 1.$$

The case $\frac{m-m_2}{2} \geq 2$ implies $r \geq 2p \geq q$ against (3.21).
The case $\frac{m-m_2}{2} = 1$ implies $r = p$ against (3.20).
The case $\frac{m-m_2}{2} = \frac{1}{2}$ implies $m_2 = m - 1$. Substituting it into (h) and using (3.23), we obtain

$$\frac{N}{N-2} = \frac{m}{m-1},$$

which is equivalent to

$$N = 2m,$$

which is impossible since by assumption N is odd.

Finally, the case $\frac{m-m_2}{2} = \frac{3}{2}$ implies $m_2 = m - 3$. Substituting it into (h) and using (3.23), we obtain

$$\frac{N}{N-2} = \frac{m}{m-3},$$

which is equivalent to

$$1 + \frac{2}{N-2} = 1 + \frac{3}{m-3},$$

whence

$$3N = 2m$$

which is also impossible since by assumption N is odd.

Therefore, the possibility $Nr = mp$ is excluded as well.

Thus a 3-tail congruent to T_3 can appear in S only at the distance f from the last partial of S.

Now let us show that $J = I$ to within a convolution unit. Consider the tail T_2 with the distance d between its two impulses. Since S contains two subspectra congruent to T_2, namely $\delta_a * T_2$ (for certain a) and $\delta_{a+f} * T_2$, and the distance between the T_2's impulses $d < f$, one of impulses of $T_2 * \delta_a$ is not superimposed on any harmonic of $T * \delta_{a+f}$. This not-superimposed impulse of $T_2 * \delta_a$ can be confronted to the corresponding impulse of $T * \delta_{a+f}$ which by Lemma 5 is also superimposed on no other impulse of lower tones. Consequently, the ratio of these "pure" harmonics determines the ratio of coefficients in I.

Thus $J = I$ to within units. By virtue of Lemma 1 and uniqueness of polynomial division, this implies $U = T$ to within a convolution unit. By Lemmas 3 and 4 factors I and T are irreducible and no further deconvolution of J and U is possible, as required. ∎

Note that this lemma is valid not only for harmonic tones with all successive partials, but, by virtue of Lemma 3, for segments of harmonic spectra which contain three higher harmonics.

Theorem 1 (Uniqueness of Interval Decomposition) *Consider a two-impulse interval distribution I and let the distance f between its impulses be smaller than or equal to 12 semitones, i.e.*

$$I = a_1 \delta_{\omega_1} + a_2 \delta_{\omega_2} \quad (\omega_1 < \omega_2);$$
$$f = \omega_2 - \omega_1 \leq 12 \text{ semitones.}$$

Let T be a harmonic spectrum with N successive partials, where N sufficiently large so that the distance between the last four harmonics of T is less than f, i.e.

$$T = \sum_{n=1}^{N} b_n \delta_{\omega_n} \quad (\omega_1 < \ldots < \omega_N);$$

$$f > \omega_N - \omega_{N-2}.$$

Consider spectrum $S = T * I$. Then its deconvolution into non-trivial factors (i.e. having at least two impulses each) is unique to within order and units.

PROOF OF THEOREM 1. By virtue of Lemma 6 it suffices to consider the case when N is even. Consider a deconvolution of S into two non-trivial factors U and J, i.e. $U * J = S$. By virtue of Lemma 5 we can assume that

$$U_4 \sim T_4.$$

Now consider U_4 without its last partial, and repeat the proof of Lemma 6 for three frequencies $(N-1)r$, $(N-2)r$, and $(N-3)r$, substituting $N' = N-1$ for N everywhere in the proof of Lemma 6. Since by assumption N is even, N' is odd, and the proof remains valid. ■

Lemma 7 (Uniqueness of Chord Decomposition) *Consider an interval distribution I with three-impulses. Let the distance between its extreme impulses be smaller than or equal to the octave and let the distances between its adjacent impulses be different, i.e.*

$$I = \sum_{i=1}^{3} a_i \delta_{\omega_i} \quad (\omega_1 < \omega_2 < \omega_3);$$

$$f = \omega_3 - \omega_1 \leq 12 \text{ semitones},$$

$$\omega_2 - \omega_1 \neq \omega_3 - \omega_2.$$

Let T be a harmonic spectrum with N successive partials, where N is not divisible by 2 and 3, and is sufficiently large so that the distance between the last four harmonics of T is less than f, i.e.

$$T = \sum_{n=1}^{N} b_n \delta_{\omega_n} \quad (\omega_1 < \ldots < \omega_N);$$

$$f > \omega_N - \omega_{N-3}.$$

Consider chord spectrum $S = T * I$. Then its deconvolution into non-trivial factors (i.e. having at least two impulses each) is unique to within order and units.

PROOF OF LEMMA 7. The proof follows the ideas of Lemma 6. The difference is that instead of two possibilities (3.22) and (3.23) one must consider the third possibility, $Nr = ls$, where s is the fundamental frequency of the middle tone of the chord, i.e. $p < s < q$, and l is a positive integer such that $l < N$. Moreover, one has to add some more cases to (a)–(d) and (e)–(h), because the frequencies nr can fall onto the harmonics of this third tone with fundamental frequency s.

Suppose that in S there exists a 4-tail congruent to T_4 which is different from the 4-tails of the chord tones. Denote its fundamental frequency by r, $r \neq p, q, s$, and consider four partials Nr, $(N-1)r$, $(N-2)r$, $(N-3)r$, which by our hypothesis are inherent in spectrum S. Since these four partials are inherent in chord spectrum S built from harmonic tones with fundamental frequencies p, q, and s, at least two of these four partials fall on harmonics of the same tone.

The case when two successive partials, say Nr and $(N-1)r$, are harmonic frequencies of the same tone of the chord is brought to contradiction in the same way as in the proof of Lemma 6.

The case when a partial and the next to next partial, say Nr and $(N-2)r$, are harmonic frequencies of the same tone of the chord is also brought to contradiction in the same way as in the proof of Lemma 6.

The only new case arises when Nr and $(N-3)r$ are harmonic frequencies of the same tone, but $(N-1)r$ and $(N-2)r$ are harmonic frequencies of two other tones of the chord.

At first suppose that Nr falls onto a harmonic of the upper tone of the chord, i.e. $Nr = kq$ and $(N-3)r = k_3q$, where $N > k > k_3$. Using the reasons similar to that from the proof of Lemma 6, we obtain

$$r = \frac{k - k_3}{3}q, \quad k - k_3 \geq 1.$$

The case $k - k_3 \geq 3$ implies $r \geq q$ against (3.21).

The case $k - k_3 = 1$ implies

$$\frac{N}{N-3} = \frac{k}{k-1},$$

which is equivalent to

$$N = 3k,$$

which is impossible since by assumption N is not divisible by 3.

The case $k - k_3 = 2$ implies

$$\frac{N}{N-3} = \frac{k}{k-2},$$

which is equivalent to

$$2N = 3k,$$

which is also impossible since by assumption N is not divisible by 3.

Now suppose that Nr falls onto a harmonic of the lower or middle tone of the chord. For example, let $Nr = ls$ and $(N-3)r = l_3 s$, where $N > l > l_3$. This implies

$$r = \frac{l - l_3}{3}s, \quad l - l_3 \geq 1.$$

The case $l - l_3 \geq 6$ implies $r \geq 2s \geq 2p \geq q$ against (3.19) and (3.21).

The case $l - l_3 = 3$ implies $r = s$ against $r \neq s$.

The case $l - l_3 = 1$ implies

$$\frac{N}{N-3} = \frac{l}{l-1},$$

which is equivalent to

$$N = 3l,$$

which is impossible since by assumption N is not divisible by 3.

The case $l - l_3 = 2$ implies

$$\frac{N}{N-3} = \frac{l}{l-2},$$

which is equivalent to

$$2N = 3l,$$

which is also impossible since by assumption N is not divisible by 3.

The case $l - l_3 = 4$ implies

$$\frac{N}{N-3} = \frac{l}{l-4},$$

which is equivalent to

$$4N = 3l,$$

which is impossible since by assumption N is not divisible by 3.

The case $l - l_3 = 5$ implies

$$\frac{N}{N-3} = \frac{l}{l-5},$$

which is equivalent to

$$5N = 3l,$$

which is also impossible since by assumption N is not divisible by 3.

Thus in S any 4-tail congruent to T_4 falls onto 4-tails of the chord tones. The remainder of the proof is similar to that of Lemma 6. ∎

Theorem 2 (Uniqueness of Chord Decomposition) *Consider a three-impulse interval distribution I. Let the distance between its extreme impulses be smaller than or equal to the octave and let the distances between its adjacent impulses be different, i.e.*

$$I = \sum_{i=1}^{3} a_i \delta_{\omega_i} \quad (\omega_1 < \omega_2 < \omega_3);$$
$$f = \omega_3 - \omega_1 \leq 12 \text{ semitones},$$
$$\omega_2 - \omega_1 \neq \omega_3 - \omega_2.$$

Let T be a harmonic spectrum with N successive partials, where N is sufficiently large so that the distance between the last seven harmonics of T is less than f, i.e.

$$T = \sum_{n=1}^{N} b_n \delta_{\omega_n} \quad (\omega_1 < \ldots < \omega_N);$$
$$f > \omega_N - \omega_{N-3}.$$

Consider chord spectrum $S = T * I$. Then its deconvolution into non-trivial factors (i.e. having at least two impulses each) is unique to within order and units.

PROOF OF THEOREM 2. The statement of the theorem follows from Lemma 7 in the same way as the statement of Theorem 1 follows from Lemma 6. For N, we substitute the greatest $N' \leq N$ which is not divisible by 2 and by 3. Since obviously $N' \geq N - 3$, and the above lemma is formulated for 4-tails, the seven partials in the tail T_7 is sufficient to adapt the lemma in order to prove the theorem. ■

Note that Theorem 2 is valid also for segments of harmonic spectra with seven successive harmonics.

Thus spectra of two-tone intervals and major or minor triads are decomposable in the unique way. Consequently, the only deconvolution of a chord spectrum, which is built from harmonic tones with a sufficient number of harmonics, reveals its generation.

Further generalizations of the Unique Decomposition Theorem to multi-note chords can be formulated by analogy with the generalization of Lemma 6 and Theorem 1 to Lemma 7 and Theorem 2, respectively.

Theorem 2 doesn't state the unique decomposition of chords if the number of partials of generative tones is small. This however can be done by directly testing each particular case from a finite number of cases.

To finish this section, we shall show that the harmonicity of tones is an important condition of the Unique Deconvolution Theorem. For arbitrary power spectra the unique deconvolution doesn't hold. By virtue of Lemma 1 this is seen from the following example proposed by Chateauneuf (1993) in personal communication.

Example 2 (No Unique Factorization of Polynomials over Positive Integers) Consider the following polynomials:

$$
\begin{aligned}
p(x) &= x^2 + 2x + 2 \\
q(x) &= x^2 - 2x + 2 \\
r(x) &= x(x^2 + 2x + 2) + 1 = (x^2 + x + 1)(x + 1),
\end{aligned}
$$

where polynomials $p(x)$, $q(x)$, x^2+x+1, and $x+1$ are irreducible over integers and, consequently, over non-negative integers as well. According to the reasons which precede (3.16), polynomial

$$
p(x)q(x) = x^4 + 4
$$

is irreducible over non-negative integers. Therefore, polynomial with non-negative integer coefficients $p(x)q(x)r(x)$ can be factored into irreducible polynomials over non-negative integers as follows

$$
p(x)q(x)r(x) = [p(x)q(x)]r(x) = (x^4 + 4)(x^2 + x + 1)(x + 1). \qquad (3.24)
$$

At the same time, polynomial $p(x)q(x)r(x)$ can be factored into polynomials over non-negative integers in a different way:

$$
\begin{aligned}
p(x)q(x)r(x) = p(x)[q(x)r(x)] &= (x^2 + 2x + 2)[x(x^4 + 4) + x^2 - 2x + 2] \\
&= (x^2 + 2x + 2)(x^5 + x^2 + 2x + 2), \quad (3.25)
\end{aligned}
$$

where the first factor, $x^2 + 2x + 2$, is irreducible, being different from all irreducible factors of factorization (3.24). Hence, there exist two different factorizations of polynomial $p(x)q(x)r(x)$ into irreducible polynomials over non-negative integers, one given by (3.24) and another given by (3.25) with a further factorization of the second term $x^5 + x^2 + 2x + 2$, if such a further factorization exists.

3.6   Causality and Optimal Data Representation

One can ask a question: Why do we perceive chords as chords but not as single sounds? From the standpoint of our consideration we can reformulate this question as follows: What are the reasons in favor of decomposing spectra instead of considering them as they are?

In order to compare different representations we refer to the criterion of least complex data representation. We shall show that the representation of a chord spectrum in a form of deconvolution is the optimal representation of the chord spectrum with regard to the amount of memory needed for the storage of the spectral data. This way we justify such a representation of spectral data and adduce reasons in favor of perceiving chords as chords but not as indivisible sounds.

Recall that according to Kolmogorov, the complexity of data is defined to be the amount of memory required for their storage (Kolmogorov 1965; Calude 1988). Since the spectra considered can be stored as a sequence of impulses, the complexity of a spectral representation can be identified with the number of impulses to be stored.

By *complexity of a spectrum* S we understand the number of points in its support Δ_S. The complexity of S is denoted by $|\Delta_S|$.

By *complexity of a deconvolution* $S = T * I$ we understand the total complexity of the factors which is equal to $|\Delta_T| + |\Delta_I|$.

Theorem 3 (Revealing Causality by Optimal Data Representation)
Suppose that a spectrum S is generated by a spectrum T translated according to an interval distribution I, where T is a harmonic spectrum or its segment with seven or more partials, and I corresponds to a two-tone interval, major triad, or minor triad. If the frequency resolution is sufficiently accurate then the spectrum representation corresponding to the spectrum generation (3.13) is the least complex representation of S.

PROOF OF THEOREM 3. By Theorem 2 there is a unique deconvolution of a chord spectrum S given by its generation (3.13). Therefore, the problem is brought to comparing the complexity $|\Delta_S|$ of the spectrum S and the complexity of its deconvolution which is

$$|\Delta_{a\delta_p}| + |\Delta_T| + |\Delta_I| = |\Delta_T| + |\Delta_I| + 1.$$

Obviously, if the translations of T according to the interval distribution I have no common partials, then

$$|\Delta_S| = |\Delta_T| \cdot |\Delta_I|.$$

Hence, if I has more than one partial, and T has more than four partials, the two-level representation of the chord spectrum is more efficient than storing the whole spectrum:

$$|\Delta_I| + |\Delta_T| + 1 < |\Delta_I| \cdot |\Delta_T| = |\Delta_S|.$$

This case however corresponds to dissonant intervals.

Consonant intervals are characterized by common partials of constituent tones. Hence,

$$|\Delta_S| < |\Delta_T| \cdot |\Delta_I|.$$

Obviously, the more partials tone spectrum T contains, the more non-coinciding partials are inherent in spectrum S. Therefore, the difference between the left-hand and right-hand parts of the above inequality is greatest when the number of partials per voice is smallest and intervals are most consonant, i.e. fifths or octaves.

It is easy to see that the spectrum of the fifth generated by a tone with seven harmonics contains 12 partials (the second harmonic of the upper tone falls on the third harmonic of the lower tone, and the fourth harmonic of the upper tone falls on the sixth harmonic of the lower tone).

The spectrum of the octave generated by a tone with seven harmonics contains 11 partials (even harmonics of the lower tone falls on the harmonics of the upper tone).

Obviously, the spectrum of a three-tone chord contains not less partials than the spectrum of its two tones. Consequently, the spectrum of a three-tone chord contains not less than 11 partials.

On the other hand, the complexity of our two-level representation of intervals chords (as a convolution of tone spectrum T and interval distribution I) for harmonic tones with seven partials is given by the estimation

$$|\Delta_T| + |\Delta_I| + 1 = 7 + 2 + 1 = 10 < 11,$$

whence we obtain the statement of the theorem for two-tone intervals.

For three-tone chords the complexity of the two-level representation is given by the estimation

$$|\Delta_T| + |\Delta_I| + 1 = 7 + 3 + 1 = 11 \leq 11,$$

whence we obtain the statement of the theorem for three-tone chords. ■

3.7   Interpretation of the Results

Theorem 3 justifies the perception of a chord as a compound sound but not as a sound entirety. Assuming that the perception performs optimal data representation, we obtain its capacity to recognize physical causality in spectral data generation. This way we get a semantical knowledge concerning the chord, using some general principles of data processing only.

The role of logarithmic scale is also remarkable. Owing to the use of logarithm, linear patterns (tone spectra with multiplication of partial frequencies) become irreducible, and their superpositions become uniquely decomposable (by deconvolution of a chord spectra). Therefore, the role of logarithmic scales in perception can be explained as contributing to pattern separation.

The role of the insensitivity of the ear to the phase of signal is remarkable either. In fact, by virtue of the isomorphism between discrete spectra and polynomials (Lemma 1), we have the unique deconvolution of discrete spectra with spectral coefficients a_n from a given number field, and also for integers, or Gaussian integers (complex numbers $a + bi$ with integer a and b). Therefore, for discrete spectra with complex coefficients we obtain the unique deconvolution theorem, corresponding to the unique factorization theorem for polynomials over the field of complex numbers. Since by the fundamental theorem of algebra a polynomial over the field of complex numbers can be factored into polynomials of the first degree, this implies the deconvolution of a discrete spectrum into discrete spectra with two adjacent impulses. (From the formal point of view this resembles the decomposition of a digital filter into a superposition of second-order filters).

Being applied to our consideration, such a result would mean that a a chord spectrum S can be factored into a convolution product of two-impulse spectra

$$S = \delta_p * I_1 * \ldots * I_n,$$

where each

$$I_i = a_i \delta_0 + b_i \delta_1 \quad (i = 1, \ldots, n)$$

is a two tone interval distribution with two closest impulses. In other words, a generative spectrum with two close impulses is translated by the minimal possible interval (according to the interval distribution with two close impulses), the resulting spectrum is translated again by the minimal possible interval, etc. We would obtain a hierarchy of "elementary blocks", minimal intervals over minimal intervals, which has nothing in common with human perception and physical causality in sound.

A similar situation arises for spectra with integral coefficients (including negative numbers). As seen from Example 1, the sensitivity of the ear to the inverse phase, corresponding to negative spectral coefficients, implies the decomposition of interval distribution of a major third into a convolution product of two distributions.

Thus, assuming the sensitivity of the ear to the phase of the signal (considering spectra with complex or integer coefficients), we obtain the impossibility to recognize chords corresponding to their generation.

Note that since the convolution is commutative with respect to its arguments,

$$S = I * T = T * I,$$

the interval distribution I can be regarded as a generative tone and T can be regarded as an interval distribution. Formally, both I and T can be considered as spectra of some signals.

It is known that the Fourier spectrum of a product of two signals equals to the convolution of the associated Fourier spectra, i.e.

$$F[s \cdot t] = F[s] * F[t], \qquad (3.26)$$

where s and t are signals as functions of time, and $F[\cdot]$ is Fourier operator. For audio spectra (with logarithmic scale of frequencies), the convolution of two spectra doesn't correspond to the product of the associated signals, but we can still say that the associated signals interact somehow. Therefore, if a chord audio spectrum $S = I * T$, the interval distribution I can be considered as a spectrum I of a certain signal which interacts with the tone signal with spectrum T.

Problem 1 *It is interesting to express analytically the interaction of the tone signal associated with spectrum T and the signal associated with the interval distribution I, i.e. to formulate a property of audio spectra analogous to (3.26).*

Since an interval distribution I can be matched to the signal $F^{-1}[I]$, the concepts of pitch and timbre are applicable to interval distributions as if they were tone spectra.

Problem 2 *It is interesting to answer whether the type of chord (major, minor, etc.) with interval distribution I corresponds to a certain "timbre" of the signal $F^{-1}[I]$ which is associated with I, and whether the chord root corresponds to the "pitch" of the signal $F^{-1}[I]$.*

The problem of determining the root note of a chord by its structure was posed as early as in the 18th century by J.-Ph. Rameau (1683–1764).

Finally, we would like to mention that the isomorphism between discrete spectra and polynomials over integers implies the applicability of polynomial division algorithms to the deconvolution of discrete spectra. However, the deconvolution method is efficient for justifying the principle of correlativity of perception theoretically, but may fail in practical applications to chord recognition. Indeed, in spectral approximations, partials can deviate not only from their values, corresponding to deviations of polynomial coefficients, but also from their frequencies, corresponding to changes of the degree of the associated polynomials. Since polynomial factorizations are very sensitive to the degree of polynomials, a slight spectral distortion may result in a considerable change of the spectral deconvolution. Consequently, spectral deconvolutions are unstable with respect to spectral distortions implying the instability of chord recognition with respect to small deviations of spectra. Therefore, the deconvolution method is not fitted well for chord recognition.

Another practical disadvantage of the deconvolution approach is that factoring real spectra requires much computing when the associated polynomials are of high degree. One can find the details of factoring polynomials by Kronecker's method in (Rédei 1967).

Therefore, our method of finding spectral representations by means of correlation analysis outlined in the previous chapter can be more stable and reliable in applications. On the other hand, note that by virtue of the isomorphism between spectra and polynomials, the correlation method may be used in order to factorize polynomials.

The unique factorization is valid for the polynomials in several variables (Birkhof & Mac Lane 1965, p. 76), as well as the Kronecker's method (van der Waerden 1953). Consequently our theoretical consideration is applicable to images which have two-dimensional spectra. However, by the same reasons, practical applications should be based on correlation analysis rather than on the factorization method.

3.8 Main Items of the Chapter

Summing up what has been said, let us formulate the main items of the chapter.

1. We have suggested a way of chord representation as generated by a tone spectral pattern translated along the $\log_2$-scaled frequency axis. It is formalized by a convolution of a tone spectral pattern and an interval distribution, corresponding to the chord structure.

2. The justification of chord recognition in its theoretical formulation is understood as the problem of unique deconvolution of the chord spectrum, corresponding to the chord spectrum generation. We have proved the indecomposability of spectra of harmonic tones and of interval distributions of two-tone intervals and major or minor triads. This corresponds to perceiving musical tones as entire sound objects and unambiguously recognizing intervals, major chords, and minor chords. Then we have established the unique deconvolution of spectra of two-tone intervals and major or minor chords with harmonic tones.

3. Note the role of logarithmic scale in our consideration. Owing to the use of logarithm, patterns with a linear structure (tone spectra with multiplication of partial frequencies) are non-linearly compressed, becoming irreducible. Therefore, the role of logarithmic scales in perception can be explained as providing the conditions for indecomposability of patterns. In particular, as follows from Lemma 3, the harmonic spectrum is irreducible, which meets the perception of a musical tone as an entire sound object rather than as a sound complex.

4. On the other hand, Theorem 3 explains the perception of chords as composed sounds. It is noteworthy that the optimality of the representation corresponds to the causality of spectral data generation. This way we obtain a semantical knowledge concerning the chord structure, using data processing only.

5. Note that if we assumed the model sensitivity to the phase of the signal, we couldn't prove the irreducibility of harmonic spectra and interval distributions of chords. On the contrary, we have shown that all spectra would become reducible to trivial two-impulse spectra. This may explain the insensitivity of audio perception to the phase of the signal: Otherwise, the signal decomposition would not correspond to physical causality in the signal generation.

Chapter 4

Implementing the Model

4.1   Reduction of the Model

In the previous chapter we have established that the generation of a chord spectrum can be uniquely reconstructed from its factorization into a convolution product of irreducible spectra. In a sense, we have proved the uniqueness theorem for chord recognition.

However, this theorem is formulated for idealized conditions. It is assumed that all the chord tones have precisely the same spectrum translated along the $\log_2$-scaled frequency axis.

As a theoretical fact, this theorem characterizes some fundamental trend in chord representation, but in reality the condition mentioned holds only approximately. Even if the tones of a chord are played on the same instrument, their spectra vary with register and pitch.

For example, consider a violin. The resonance characteristic of the violin body is constant, say, dumping f_1 and enhancing a_1. This implies the first partial of tone f_1 to be less salient in the tone spectrum than the first partial of tone a_1. Consequently, the spectrum of tone a_1 is not equal to the spectrum of tone f_1 translated by the major third along the frequency axis $\log_2$-scaled.

The above reason makes difficult direct applications of the formalism from the previous chapter. In its present form it is applicable to the recognition of structure based on identity (replications of the same tone spectrum). However, in reality we deal with the recognition of structure based on similarity (replications of similar tone spectra).

In order to overcome this difficulty, we reduce our consideration to clipped spectra, or Boolean spectra whose partials have amplitudes 0 or 1. This way the attention is restricted to spectral structures which are determined by the ratio of partial frequencies but not by partial amplitudes, or spectral envelopes.

Boolean spectra are much more stable with respect to pitch translations than the original power spectra. Indeed, in the above example concerning violin sounds, the amplitude of tone partials varies, but their frequency ratio is always constant being determined by the harmonic ratio

$$1 : 2 : 3 : \ldots : k : \ldots .$$

In particular, this implies that the Boolean spectrum of violin tone a_1 can be considered as almost identical to the Boolean spectrum of tone f_1 translated by a major third along the $\log_2$-scaled frequency axis.

Therefore, we formulate the problem of chord recognition for Boolean spectra, restricting our attention to the frequency structure and ignoring spectral envelopes. In our case, the recognition of acoustical structure can fail if the structure is understood as based on identity but not on similarity.

Here we propose a kind of a compromise. We do not consider the similarity with deviations of all the characteristics of tone spectra but only that with deviations of partial intensities. In a general model, the deviations of partial frequencies should be also taken into account, but fortunately the harmonic ratio of partial frequencies is quite stable in musical voices. Thus, owing to this property of musical tones, the recognition of similarity in chord spectra is replaced by the recognition of identity in their Boolean spectra.

The recognition of structure based on similarity is not as unambiguous as the recognition of structure based on identity. Indeed, we show that unlike power spectra the unique deconvolution is not valid for Boolean spectra. In particular, this means that the uniqueness theorem for chord recognition is not true. However, the ambiguity in spectral representation arises quite seldom, and we can provide measures in order to control such situations. On the other hand, the recognition of chords based on Boolean spectra has evident advantages: It is much simpler and faster.

Obviously, if a spectrum can be factored into a convolution product, the same can be done for the corresponding Boolean spectrum. Consequently, the Boolean reduction of our model provides for some necessary but not sufficient means for chord recognition. In a sense, recognizing chords by their Boolean representations is somewhat similar to finding the minimum of a function by the points where the first derivative of the function is equal to zero: This condition doesn't guarantee the solution, but the search is reduced to a few points.

In Section 4.2, "Properties of Boolean Spectra", we prove that the chord representation by their Boolean spectra has the same properties as the chord representation by discrete power spectra, except for the unique deconvolution property. The ambiguity in the deconvolution of a chord spectrum is illustrated by a simple example.

In Section 4.3, "Necessary Condition for Generative Patterns", we show how autocorrelation analysis can be used for finding generative patterns in chord spectra. Then we extend simple autocorrelation analysis to multiauto-correlation analysis with which multiple replications of a generative pattern can be detected. Finally, we formulate a theorem on necessary conditions for generative patterns and their search in terms of recurrent autocorrelation analysis.

In Section 4.4, "Algorithm for Finding Generative Patterns", the procedure of chord recognition based on the theorem from the previous section is traced step by step. This procedure uses a series of embedded cycles of simple autocorrelation analysis of a chord spectrum. It is explained that in spite of embedded cycles the branching of this procedure is rather limited.

In Section 4.5, "Summary of Reduced Model", the main items of the chapter are summarized.

4.2 Properties of Boolean Spectra

As said in the previous section, we shall restrict our consideration to Boolean spectra.

Let

$$S = S(x) = \sum_{n=0}^{N} S_n \delta(x - n) = \sum_{n=0}^{N} S_n \delta_n$$

be a discrete spectrum with non-negative integer coefficients S_n as defined in (3.7).

The *Boolean spectrum* (associated with S) is defined to be the expression

$$s = \bigvee_n s(n)\delta_n. \tag{4.1}$$

where

$$s(n) = \begin{cases} 0 & \text{if } S_n = 0, \\ 1 & \text{if } S_n \neq 0. \end{cases}$$

(The designation s_n instead of $s(n)$ might seem to be more natural, but in the sequel we shall use indexes for other purposes).

One can compare the power spectra in Fig. 3.2–3.3 with their associated Boolean spectra in Fig. 4.1–4.2.

Since we consider the $\log_2$-scaled frequency axis, pitch transpositions correspond to parallel translations of the tone spectrum along the frequency axis. A translation of a Boolean spectrum (4.1) by m frequency bands to the right corresponds to the convolution

$$\delta_m * s = \bigvee_n s(n)\delta_{n+m} = \bigvee_n s(n - m)\delta_n. \tag{4.2}$$

Similarly, a multiple translation of a Boolean spectrum s is determined by the convolution of s and another Boolean spectrum,

$$i = \bigvee_n i(n)\delta_n,$$

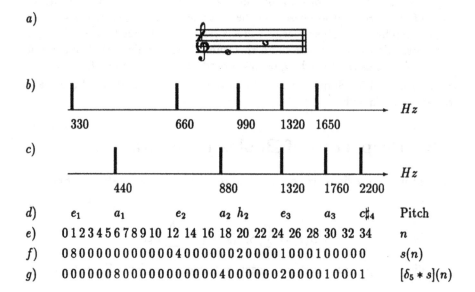

a)

b)

c)

| d) | e_1 | a_1 | e_2 | a_2 h_2 | e_3 | a_3 | $c\sharp_4$ | Pitch |

e) 0 1 2 3 4 5 6 7 8 9 10 12 14 16 18 20 22 24 26 28 30 32 34 n

f) 0 8 0 0 0 0 0 0 0 0 0 0 0 0 4 0 0 0 0 0 0 2 0 0 0 1 0 0 0 1 0 0 0 0 0 $s(n)$

g) 0 0 0 0 0 0 8 0 0 0 0 0 0 0 0 0 0 0 4 0 0 0 0 0 0 2 0 0 0 1 0 0 0 1 $[\delta_5 * s](n)$

Figure 4.1: Example of Boolean spectra of tones

a) tones e_1 and a_1 in standard musical notation;

b) the Boolean spectrum (with $\log_2$-scaled frequency axis) of tone e_1 for a harmonic voice with 5 successive partials which have equal power;

c) the same for tone a_1;

d) the pitches $e_1, \ldots, c\sharp_4$ which correspond to the mean frequencies of the frequency bands for the given spectral representations;

e) the indexes $n = 1, \ldots, 34$ of frequency bands for the given discrete spectra;

f) the Boolean spectrum of tone e_1 under the frequency resolution within a semitone (1/12 octave) — string $s(n)$;

g) the same for the second tone — string $[\delta_5 * s](n)$.

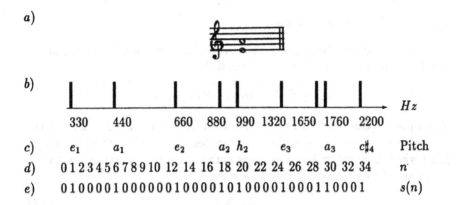

a) chord (e_1, a_1) in standard musical notation;

b) the Boolean spectrum (with $\log_2$-scaled frequency axis) of the chord for a harmonic voices with 5 successive partials which have equal power;

c) the pitches $e_1, \ldots, c\sharp_4$ which correspond to the mean frequencies of the frequency bands for the given spectral representations;

d) the indexes $n = 1, \ldots, 34$ of frequency bands for the given discrete spectra;

e) the Boolean spectrum of the chord under the frequency resolution within a semitone (1/12 octave) — string $s(n)$.

which is called the *interval distribution*.

The Boolean interval distribution associated with the interval of m semi-tones is defined to be

$$i_m = \delta_0 \vee \delta_{[mC/12+0.5]}, \tag{4.3}$$

where the denotations are the same as in (3.8):

C is the constant which characterizes the accuracy of spectral representation, being equal to the number of frequency bands per octave,

$[\cdot + 0.5]$ is the rounding function.

The interval distribution of a three-tone chord with intervals of m and k semitones from its lowest tone is defined to be

$$i_{m,k} = \delta_0 \vee \delta_{[mC/12+0.5]} \vee \delta_{[kC/12+0.5]}.$$

Example 3 (Boolean Interval Distribution of Major Third) Let the frequency resolution be to within one semitone ($C = 12$). Then the major third which is equal to four semitones corresponds to four frequency bands. Consequently, its interval distribution is equal to

$$i_4 = \delta_0 \vee \delta_4. \tag{4.4}$$

Example 4 (Boolean Interval Distribution of Major Triad) Let the frequency resolution be to within quarter of a tone ($C = 24$). The major triad in root position is determined by intervals of four and seven semitones from the lowest tone, which correspond to 8 and 14 frequency bands, respectively. Consequently, the interval distribution of this major triad is equal to

$$i_{4,7} = \delta_0 \vee \delta_8 \vee \delta_{14}. \tag{4.5}$$

As follows from (3.9), the Boolean harmonic spectrum has the form

$$s = \bigvee_{k=1}^{K} \delta_{p+[C\log_2 k+0.5]} = \delta_p * \bigvee_{k=1}^{K} \delta_{[C\log_2 k+0.5]}.$$

Example 5 (Boolean Harmonic Spectrum) Let the frequency resolution be to within one semitone ($C = 12$). Suppose that index $n = 0$ corresponds to the frequency band centered at $262Hz$ which corresponds to the note c_1 (*do* of the first octave). Consider a musical tone a_1 (*la* of the first octave) with five successive harmonics. Then the frequency band corresponding to a_1 has the index $n = 9$. Hence, the Boolean spectrum of our tone a_1 is equal to

$$s = \delta_9 \vee \delta_{21} \vee \delta_{28} \vee \delta_{33} \vee \delta_{37} \tag{4.6}$$

$$= \delta_9 * [\delta_0 \vee \delta_{12} \vee \delta_{19} \vee \delta_{24} \vee \delta_{28}]. \tag{4.7}$$

A Boolean spectrum s is said to be *simple* if its first coefficient $s(0) \neq 0$.

For example, the interval distributions (4.4–4.5) are simple. On the contrary, the Boolean spectrum (4.6) is not simple. In (4.7) the same spectrum as in (4.6) is represented as a simple Boolean spectrum translated by 9 bands to the right.

Since all harmonic spectra are similar in their structure, a chord (even with pairwise different harmonic voices) can be approximately regarded as generated by translations of a harmonic voice pattern along the frequency axis. The inequality of voice spectra may result in small errors in such a representation. Hence, we formulate the following basic assumption.

Conjecture 10 (Representation of Chord Spectra) *The Boolean spectrum of a chord can be approximately represented as generated by multiple translations of a tone spectral pattern as follows*

$$s = \delta_p * t * i + \epsilon - \lambda, \tag{4.8}$$

where

p *is the index of the first frequency band of the lowest tone of the chord (conventional pitch of the chord's lowest tone),*

t *is a tone pattern which is a simple Boolean spectrum,*

i *is an interval distribution which is a simple Boolean spectrum,*

ϵ *is the set (spectrum) of missed partials which should be added to $\delta_p * t * i$ in order to obtain s,*

λ *is the spectrum (spectrum) of false partials which should be removed from $\delta_p * t * i$ in order to obtain s.*

We suppose that the error-correcting spectra ϵ and λ in (4.8) contain rather few partials so that the term

$$s' = \delta_p * t * i$$

approximates the Boolean spectrum s. Therefore, the deconvolution properties of the Boolean spectrum s' are of prime importance.

In the previous chapter it is shown that a chord spectrum with non-negative integral coefficients is uniquely factored into the convolution product of irreducible spectra with non-negative integral coefficients, while harmonic spectra and interval distributions of two-tone intervals and major or minor triads being irreducible. In case of Boolean spectra the situation is different. As seen from the following example, Boolean spectra are not uniquely factored into the convolution product of irreducible Boolean spectra, making the deconvolution problem ambiguous.

Example 6 (No Unique Deconvolution of Boolean Spectra) Let the
frequency resolution be to within one semitone ($C = 12$). Suppose that index
$n = 0$ corresponds to the frequency band centered at $262Hz$ which corresponds
to note c_1 (*do* of the first octave). Consider musical tones c_1 and c_2 (*do* of
the first octave and *do* of the second octave which are shown in Fig. 4.3a) and
two-tone interval $(c_1; c_2)$.

At first suppose that the voices have four successive harmonics. Their
spectra are shown in Fig. 4.3b–c. To be precise, we consider the following
Boolean tone spectrum

$$t = \delta_0 \vee \delta_{12} \vee \delta_{19} \vee \delta_{24}. \tag{4.9}$$

and the Boolean interval distribution

$$i = \delta_0 \vee \delta_{12}. \tag{4.10}$$

The convolution of these two spectra, corresponding to the chord spectrum of
the interval $(c_1; c_2)$, equals to

$$s = t * i = \delta_0 \vee \delta_{12} \vee \delta_{19} \vee \delta_{24} \vee \delta_{31} \vee \delta_{36}.$$

This spectrum is shown in the third line of Fig. 4.3b, and in its binary form—in
the third line of Fig. 4.3c.

Now omit the second partial of tone t and consider the resulting Boolean
spectrum

$$t' = \delta_0 \vee \delta_{19} \vee \delta_{24}. \tag{4.11}$$

This spectrum and its translation by the octave are shown in Fig. 4.3d–e. One
can see that

$$s = i * t = i * t'.$$

It is illustrated by the fact that the third line in Fig. 4.3b is equal to the third
line in Fig. 4.3d and the third line in Fig. 4.3c is equal to the third line in Fig.
4.3e.

Note that Lemma 2 is formulated in terms of the properties of spectral supports. This implies its applicability to Boolean spectra as well. Consequently,
by virtue of Lemma 2 the Boolean spectra t, t', and i are irreducible, implying the ambiguity of the deconvolution of Boolean spectrum s into irreducible
Boolean spectra.

If we considered power spectra, the ambiguity in the spectral deconvolution
wouldn't occur. Indeed, similarly to (4.9–4.11) put

$$
\begin{aligned}
T &= \delta_0 + \delta_{12} + \delta_{19} + \delta_{24}, \\
I &= \delta_0 + \delta_{12}, \\
T' &= \delta_0 + \delta_{19} + \delta_{24}
\end{aligned}
$$

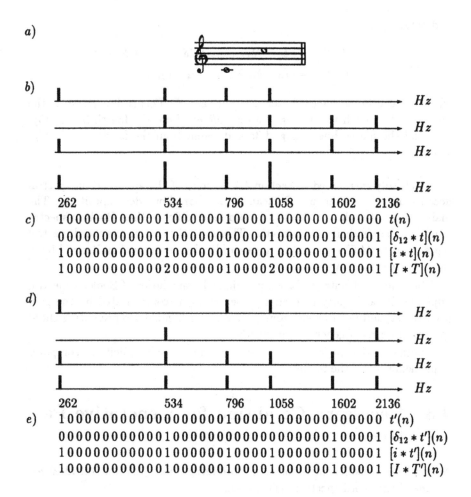

a)

b)

262   534   796 1058   1602 2136

c)
```
1 0 0 0 0 0 0 0 0 0 0 0 1 0 0 0 0 0 1 0 0 0 1 0 0 0 0 0 0 0 0 0 0 0   t(n)
0 0 0 0 0 0 0 0 0 0 0 0 1 0 0 0 0 0 0 0 0 0 1 0 0 0 0 0 1 0 0 0 0 1   [δ₁₂ * t](n)
1 0 0 0 0 0 0 0 0 0 0 0 1 0 0 0 0 0 1 0 0 0 1 0 0 0 0 0 1 0 0 0 0 1   [i * t](n)
1 0 0 0 0 0 0 0 0 0 0 0 2 0 0 0 0 0 1 0 0 0 2 0 0 0 0 0 1 0 0 0 0 1   [I * T](n)
```

d)

262   534   796 1058   1602 2136

e)
```
1 0 0 0 0 0 0 0 0 0 0 0 0 0 0 0 0 0 1 0 0 0 1 0 0 0 0 0 0 0 0 0 0 0   t'(n)
0 0 0 0 0 0 0 0 0 0 0 0 1 0 0 0 0 0 0 0 0 0 0 0 0 0 0 0 1 0 0 0 0 1   [δ₁₂ * t'](n)
1 0 0 0 0 0 0 0 0 0 0 0 1 0 0 0 0 0 1 0 0 0 1 0 0 0 0 0 1 0 0 0 0 1   [i * t'](n)
1 0 0 0 0 0 0 0 0 0 0 0 1 0 0 0 0 0 1 0 0 0 1 0 0 0 0 0 1 0 0 0 0 1   [I * T'](n)
```

Figure 4.3: No unique deconvolution of Boolean spectra

a) tones c_1 and c_2 in standard musical notation;

b) the tone spectra for voices with four successive harmonics, their Boolean union, and their sum;

c) the same in the Boolean string form;

d) the tone spectra for voices with first, third, and fourth harmonics, their Boolean union, and their sum;

e) the same in the Boolean string form.

and obtain

$$
\begin{aligned}
S = I * T &= \delta_0 + 2\delta_{12} + \delta_{19} + 2\delta_{24} + \delta_{31} + \delta_{36}, \\
I * T' &= \delta_0 + \delta_{12} + \delta_{19} + \delta_{24} + \delta_{31} + \delta_{36},
\end{aligned}
$$

whence S can be no longer factored into T' and I. This is illustrated by the fact that the fourth line in Fig. 4.3b is different from the fourth line in Fig. 4.3d and the fourth line in Fig. 4.3c is different from the fourth line in Fig. 4.3e.

As already mentioned in Section 4.1, in spite of the ambiguity in spectral deconvolution, we shall use Boolean spectra for chord decomposition. The main argument is the fact that the chord tones differ in their Boolean spectra much less than in their power spectra. Therefore, the error in the chord spectral representation (4.8) is much less in case of Boolean spectra than it could be in case of power spectra.

Thus the disadvantage of the non-unique deconvolution of Boolean spectra, implying the ambiguity in their representation, is compensated by the gain from the stability of Boolean tone patterns which enables representing chord spectra in terms of generative elements.

Another reason in favor of Boolean spectra is the simplicity of computer implementation of the model.

4.3 Necessary Condition for Generative Patterns

At first let us consider the problem of chord recognition, assuming the idealized scheme of the chord spectrum generation.

Suppose that a Boolean chord spectrum s in (4.8) is generated without variations of the tone pattern t while translating it along the frequency axis. Obviously, in this case no error-correcting terms ϵ and λ are needed. In other words, let

$$
s = \bigvee_n s(n)\delta_n = \delta_p * t * i, \tag{4.12}
$$

where

$$
\begin{aligned}
t &= \bigvee_n t(n)\delta_n, \\
i &= \bigvee_n i(n)\delta_n
\end{aligned}
$$

are simple Boolean spectra. The question is: How can we find the original deconvolution (4.12)?

In order to find repetitive subspectra in s, one can analyze peaks of the autocorrelation function of the Boolean chord spectrum

$$R(x) = \sum_{n=x}^{N} s(n)s(n-x). \tag{4.13}$$

Obviously, if there is an interval of j frequency bands in the interval distribution i, i.e. if $i(j) = 1$, implying the translation of tone pattern t by j positions to the right, then the autocorrelation function $R(x)$ has a peak at the point $x = j$. The *correlated group of partials* (spectrum)

$$t_j = \bigvee_{n:t(n)t(n+j)\neq 0} \delta_n$$

besides the partials from the generative spectrum can have some accidental partials, so that we have

$$t \subset t_j.$$

Similarly, if there are intervals $j_1, \ldots, j_k$ in the interval distribution i then the *multiautocorrelation function*

$$R(x_1, \ldots, x_k) = \sum_{n=\max\{x_1,\ldots,x_k\}}^{N} s(n)s(n-x_1)\cdots s(n-x_k) \tag{4.14}$$

has a peak at the point $(x_1, \ldots, x_k) = (j_1, \ldots, j_k)$. The *multicorrelated group of partials* (spectrum)

$$t_{j_1,\ldots,j_k} = \bigvee_{n:t(n)t(n+j_1)\ldots t(n+j_k)\neq 0} \delta_n$$

has all the partials from the generative tone spectrum t and may have some accidentals, so that

$$t \subset t_{j_1,\ldots,j_k}.$$

Since we consider Boolean spectra, the value of the autocorrelation function of a spectrum is equal to the number of coinciding ones in the spectrum and its copy translated by a corresponding number of frequency bands. Similarly, the value of the multicorrelation function of a spectrum is equal to the number of coinciding ones in the spectrum and its copies translated by corresponding numbers of frequency bands. Obviously, if we reduce the dimension of multiautocorrelation, meaning that one of translations of the spectrum is not taken into account, the number of coinciding ones increases or remains the same, implying increase or equality in the value of the multiautocorrelation function. Therefore, for arbitrary values of arguments $x_1, \ldots, x_n$ it holds

$$R(x_1, \ldots, x_k) \leq R(x_1, \ldots, x_{k-1}) \leq \cdots \leq R(x_1).$$

Hence, we obtain the following theorem.

Theorem 4 (Necessary Condition of Generative Tone Pattern) *Let a Boolean spectrum s of a chord be generated by a tone pattern t translated according to an interval distribution i, i.e. let*

$$s = t * i, \qquad (4.15)$$

where

$$i_{j_1,\dots,j_k} = \delta_0 \vee \delta_{j_1} \vee \dots \vee \delta_{j_k} \qquad (j_1 < \dots < j_k). \qquad (4.16)$$

Then

- *the multiautocorrelation function $R(x_1,\dots,x_k)$ has a peak at the point $(j_1,\dots,j_k)$,*

- *the multiautocorrelation function $R(x_1,\dots,x_{k-1})$ has a peak at the point $(j_1,\dots,j_{k-1})$,*

- *...,*

- *the autocorrelation function $R(x_1)$ has a peak at the point j_1, and*

- *the associated correlated subspectra contain the generative subspectrum t and are embedded as follows*

$$t \subset t_{j_1,\dots,j_k} \subset t_{j_1,\dots,j_{k-1}} \subset \dots \subset t_{j_1,j_2} \subset t_{j_1}. \qquad (4.17)$$

Thus in order to find a repeating subspectrum one should scan over the peaks of the multiautocorrelation function $R(x_1,\dots,x_k)$. These peaks can be found recurrently by peaks of the multiautocorrelation function of a smaller dimension. Moreover, by virtue of (4.16) the search for peaks of the multiautocorrelation function is reduced to the simplex of argument values which has the form

$$\{(x_1,\dots,x_k) : x_1 < \dots < x_k\}.$$

In addition to the directional search for deconvolutions (4.12), the above theorem enables finding multicorrelated patterns, while not knowing the multiplicity of the correlation. In other words, one can recognize a chord not knowing the number of its notes.

First, one should examine two-tone representations, finding most correlated groups of partials and for each group trying to reconstruct the chord spectrum from the translation of this group by the corresponding interval of correlation. If two-tone representations are not successful, most correlated groups of partials should be tested on their third appearance in the chord spectrum. For this purpose, instead of triple-correlation analysis, on has to look for the correlation between correlated group of partials previously deternmined. Then each triple-correlated group of partials should be tested as a generative pattern of the chord spectrum by trying to reconstruct the chord spectrum from the translations of this group by the corresponding intervals of correlation. If tree-tone representations are also not successful, the procedure should be continued for four-note representation, etc., by virtue of (4.17) each time being based on the most correlated groups of partials selected at the previous step.

Now suppose that the chord tones have slightly different Boolean spectra, i.e. the chord Boolean spectrum has the form (4.8) rather than the form (4.15). In such a case, instead of obtaining the precise decomposition (4.15) one obtains several alternative representations of the form (4.8). The desired representation can be chosen with respect to the criterion of least complexity.

Since the spectra considered can be stored as a sequence of impulses, the complexity of a spectral representation can be identified with the number of impulses to be stored.

By *complexity of a Boolean spectrum s* we understand the number of points in its support Δ_s. The complexity of s is denoted by $|\Delta_s|$.

By *complexity of the deconvolution $s = t * i$* we understand the total complexity of the convolution factors which is equal to $|\Delta_t| + |\Delta_i|$.

The detailed algorithm for finding generative Boolean spectra is given in the next section.

4.4   Algorithm for Finding Generative Patterns

The theorem from the previous section implies an algorithm for finding generative Boolean spectra of chords. Although the theorem is formulated for spectra of the form (4.15), the algorithm is applicable to spectra of the form (4.8), i.e. to chords whose tones have slightly different Boolean spectra, which corresponds to real situations.

Let s be a Boolean spectrum of a chord. In order to find a generative tone pattern t and an interval distribution i, one should do the following.

1. Without loss of generality we can suppose that s is simple, otherwise s can be brought to the origin by an appropriate translation $\delta_{-p} * (\cdot)$, where p is the index of the frequency band with the spectrum's lowest partial tone.

2. Perform ordinary autocorrelation analysis of Boolean spectrum s by analyzing peaks of the autocorrelation function $R(x_1)$. Let j_1 provide the function $R(x_1)$ with the maximum, j_2 provide its second maximal value, j_3 the third, etc. Consider, say, the first 10 peaks $j_1, \ldots, j_{10}$ of $R(x_1)$.

   (a) Consider the interval j_1, and the associated correlated group of partials t_1. Define the interval distribution

$$i_1 = i_{j_1} = \delta_0 \vee \delta_{j_1}.$$

Try to represent the Boolean spectrum s as generated by t_1 and i_1, putting

$$s_1' = t_1 * i_1,$$

and comparing s_1' with s. Hence, we obtain the representation

$$s = s_1' + \epsilon_1 - \lambda_1,$$

where the subspectrum ϵ_1 consists of the partials of s which are missed in s_1', and λ_1 consists of the partials of s_1' which are not present in s.

(b) Consider the interval j_2, and the associated correlated group of partials t_2. Perform all the operations of the preceding item (substituting index 2 for index 1) in order to verify whether s is generated by the correlated tone pattern translated by the interval of j_2 units.

(c) Repeat the above procedure several times (e.g. 10). Even if we fail in representing s as a two-tone interval, by virtue of Theorem 4 if the number of processed peaks of $R(x)$ is sufficiently large, we can be sure that the true tone pattern belongs to one of the spectra t_k correlated.

3. Perform triple-autocorrelation analysis of Boolean spectrum s by analyzing peaks of autocorrelation function $R(x_1, x_2)$. By virtue of Theorem 4 one should scan not all possible pairs (x_1, x_2), but only such that

- $x_1 = j_1, \ldots, j_{10}$, where $j_1, \ldots, j_{10}$ are selected at the previous step, and

- $x_2 > x_1$.

This means that we are looking for triple correlated groups of partials among correlated groups of partials.

(a) Consider the interval j_1, and the associated correlated group of partials t_1. Consider the peaks of

$$R(j_1, x_2) = \sum_n t_1(n)s(n - x_2)$$

for $x_2 > j_1$. Let $j_{1,1}$ provide the function $R(j_1, x_2)$ with the maximum, $j_{1,2}$ provide its second maximal value, etc. Consider, say the first 10 peaks $j_{1,1}, \ldots, j_{1,10}$ of $R(j_1, x_2)$.

i. Consider the intervals j_1 and $j_{1,1}$. Let the associated triple-correlated group of partials be

$$t_{1,1} \subset t_1.$$

Define the interval distribution

$$i_{1,1} = i_{j_1, j_{1,1}} = \delta_0 \vee \delta_{j_1} \vee \delta_{j_{1,1}}.$$

Try to represent s as generated by $t_{1,1}$ and $i_{1,1}$, putting

$$s'_{1,1} = t_{1,1} * i_{1,1},$$

and comparing $s'_{1,1}$ with s. Hence, we obtain the representation

$$s = s'_{1,1} + \epsilon_{1,1} - \lambda_{1,1},$$

where the subspectrum $\epsilon_{1,1}$ consists of the partials of s which are missed in $s'_{1,1}$, and $\lambda_{1,1}$ consists of the partials of $s'_{1,1}$ which are not present in s.

ii. Consider the intervals j_1 and $j_{1,2}$. Similarly, define the triple-correlated group of partials

$$t_{1,2} \subset t_1.$$

Define the interval distribution

$$i_{1,2} = i_{j_1, j_{1,2}} = \delta_0 \vee \delta_{j_1} \vee \delta_{j_{1,2}}.$$

Try to represent s as generated by $t_{1,2}$ and $i_{1,2}$, putting

$$s'_{1,2} = t_{1,2} * i_{1,2},$$

and comparing $s'_{1,2}$ with s.

iii. Repeat the above procedure several times (e.g. 10).

(b) Consider the interval j_2, and the associated correlated group of partials t_2. Consider the peaks of

$$R(j_2, x_2) = \sum_n t_2(n) s(n - x_2)$$

for $x_2 > j_2$. Let $j_{2,1}$ provide the function $R(j_2, x_2)$ with the maximum, $j_{2,2}$ provide its second maximal value, etc. Consider, say, the first 10 peaks $j_{2,1}, \ldots, j_{2,10}$ of $R(j_2, x_2)$.

   i. Consider the intervals j_2 and $j_{2,1}$. Let the associated triple-correlated group of partials be

$$t_{2,1} \subset t_2.$$

Define the interval distribution

$$i_{2,1} = i_{j_2,j_{2,1}} = \delta_0 \vee \delta_{j_2} \vee \delta_{j_{2,1}}.$$

Try to represent s as generated by $t_{2,1}$ and $i_{2,1}$, putting

$$s'_{2,1} = t_{2,1} * i_{2,1},$$

and comparing $s'_{2,1}$ with s. Hence, we obtain the representation

$$s = s'_{2,1} + \epsilon_{2,1} - \lambda_{2,1},$$

where the subspectrum $\epsilon_{1,1}$ consists of the partials of s which are missed in $s'_{2,1}$, and $\lambda_{1,1}$ consists of the partials of $s'_{2,1}$ which are not present in s.

  ii. Consider the intervals j_2 and $j_{2,2}$. Similarly, define the triple-correlated group of partials

$$t_{2,2} \subset t_2.$$

Define the interval distribution

$$i_{2,2} = i_{j_2,j_{2,2}} = \delta_0 \vee \delta_{j_2} \vee \delta_{j_{2,2}}.$$

Try to represent s as generated by $t_{2,2}$ and $i_{2,2}$, putting

$$s'_{2,2} = t_{2,2} * i_{2,2},$$

and comparing $s'_{2,2}$ with s.

  iii. Repeat the above procedure several times (e.g. 10).

 (c) For every $j_3, \ldots, j_{10}$ perform the same cycle.

4. Perform quadruple-autocorrelation analysis of the Boolean spectrum s by finding and analyzing peaks of the autocorrelation function $R(x_1, x_2, x_3)$. By Theorem 4 one should scan not all possible triplets (x_1, x_2, x_3), but only such that

- $(x_1, x_2) = (j_k, j_{k,m})$, where j_k are selected at the first step, and $j_{k,m}$ are selected at the second step of the algorithm, and

- $x_3 > x_2$.

This means that we are looking for quadruple correlated groups of partials among triple-correlated groups of partials. For this purpose one should perform three embedded cycles, on $k = 1, \ldots, 10$ (the first interval of the chord), on $m = 1, \ldots, 10$ (the second interval of the chord depending on the first interval), and on $l = 1, \ldots, 10$, which corresponds to revealing third interval depending on the two intervals already known.

5. The search can be continued, each time having one more embedded cycle.

Note that it is not necessary to perform a fixed number of embedded cycles. Obviously, most branches of our branching process are exhausted at the very first steps of the algorithm. For instance, suppose that there is a simply (double) correlated groups of partials but the chord spectrum cannot be reconstructed from this group of partials translated by the interval of correlation, i.e.

$$s' = t' * i' \not\approx s.$$

If at the next step of the algorithm it turns out that this group of partials is not triple correlated (or the triple correlation is insufficient), then, obviously, this group cannot be quadriple correlated, etc. In this case this group of partials should be rejected from further analysis, meaning that the related branch of the algorithm is exhausted.

If the original Boolean spectrum of our chord has the form (4.15) we will obtain one or several precise deconvolutions of the form

$$s = t' * i'$$

(recall that by assumption our spectrum is simple, i.e. $p = 0$), where the error-correcting spectra ϵ and λ with appropriate indexes are zero spectra. However, owing to accidental correlations of partials, the tone spectrum t' may have more partials than t, i.e. $t \subset t'$. Therefore, one should reduce the spectrum t', trying various $t'' \subset t'$ and $i'' \subset i'$, remaining within the above representation, i.e. so that $s = t'' * i''$.

If the original Boolean spectrum of the chord has the form (4.8), one has to compare the representations

$$s = t' * i' + \epsilon' - \lambda'$$

obtained by the algorithm, trying to reduce them as described above and minimizing their complexity

$$|\Delta_{t'}| + |\Delta_{i'}| + |\Delta_{\epsilon'}| + |\Delta_{\lambda'}|.$$

This way the optimal representation of the chord spectrum can be found out. Note that it may differ from the chord generation, but such cases occur quite seldom.

4.5   Summary of Reduced Model

Let us summarize the main items of the present chapter.

1. In order to make tone spectra (almost) invariant with respect to pitch translations, it is proposed to consider Boolean spectra instead of power spectra. This enables using the model of correlative perception, where chords are understood as contours generated by translations of the tone spectrum along the $\log_2$-scaled frequency axis.

2. The properties of Boolean spectra are similar to that of power spectra, with the only exception for the unique spectral deconvolution into irreducible spectra. Reconstructing the chord generation is still possible by decomposing the Boolean spectrum of a chord into irreducible spectra. However, such a decomposition is no longer a necessary and sufficient for reconstructing the spectrum generation, but only a necessary condition.

3. We formulate a necessary condition for the decomposition of the Boolean spectrum of a chord with a reference to the multiautocorrelation function of the spectrum. The theorem proved implies a recurrent procedure of a directional search for the decomposition of a Boolean chord spectrum which corresponds to the spectrum generation.

4. We propose an algorithm for finding generative tone spectra in a Boolean spectrum of a chord. This algorithm is applicable to recognizing chords even if the Boolean tone spectra of the chord are not identical to each other. Therefore, this algorithm of recognizing acoustical structure by similarity (not only by identity) is adapted for practical purposes.

Chapter 5

Experiments on Chord Recognition

5.1 Goals of Computer Experiments

In order to investigate the properties of the model from the previous chapters, we have performed a series of computer experiments on chord recognition with synthesized data.

First of all we have implemented the following simple correlation approach.

- **Basic (Simple Correlation) Approach**

    Recall that the recognition of chords in our model is based on recognizing interval relationships between repetitive subspectra. We distinguish between two types of intervals: *harmonic*, or *vertical intervals* between tones of the same chord, and *melodic*, or *horizontal intervals* between tones of different chords. A chord is understood to be an acoustical contour which is drawn by a tone spectral pattern in the frequency domain. Since this contour is constituted by harmonic intervals, the recognition of harmonic intervals is fundamental in the recognition of separate chords. Similarly, a polyphonic part is understood to be a dynamical trajectory which is drawn by a tone spectral pattern in the frequency domain versus time. Since this trajectory is constituted by melodic intervals between every two chords, the recognition of melodic intervals is fundamental in tracking polyphonic voices.

    Harmonic intervals are found by the autocorrelation of the chord spectra with the $\log_2$-scaled frequency axis, and melodic intervals are found by the correlation of spectra of the chords confronted. The correlated groups of partials are interpreted as tone patterns. Therefore, every chord can be recognized either being taken separately, by recognizing harmonic intervals, or being confronted to another chord, by melodic intervals.

The recognition of chords by finding either harmonic or melodic intervals by correlation analysis of chord spectra constitutes the basic element of our model.

A particular difficulty in the recognition of both separate chords and chord progressions is preventing "missed" tones and rejecting "false" tones. Usually, the spectrum of a missed tone is contained in some larger correlated group of partials which is interpreted as another tone. False tones result from accidental correlations caused by structural coincidences in the chord spectra. To a great extent, this difficulty can be overcome by complementing the basic approach by the further improvements.

- **Extensive (Decision Making) Approach**

  Since each chord is recognized in several independent ways, being taken separately and being confronted to its neighbors, the results of these different recognition procedures are processed by a decision making model.

  At first, for every recognition procedure we formulate a hypothesis on the compound of the chord. For this purpose, each correlated group of partials from the intervals found is represented by "conventional pitch" (e.g. by the lowest partial from the group of partials correlated), and the chord is identified with certain "notes". Next, the final decision about accepting or rejecting a note is made with taking into account the frequency of this note in these hypotheses.

  In our experiments we have formulated three hypotheses concerning the compound of each chord, obtained from the analysis of the chord's harmonic intervals, from the analysis of melodic intervals between the chord and its predecessor, and from the analysis of melodic intervals between the chord and its successor. One can consider more hypotheses by confronting the chord to all other chords in the given progression, but even the three hypotheses mentioned are sufficient for providing a considerable improvement in the recognition reliability.

  This approach is extensive, since it is based on multihypotheses decision making by taking into account as much information about the chord in the context as possible.

- **Intensive (Structural) Approach**

  In order to reveal missed tones and recognize accidental correlations in the chord spectrum, the structure of correlated groups of partials is analyzed. Since accidental correlations arise randomly, the structure of accidentally correlated groups of partials is also random, whereas the spectral structure of true tones is stable. This observation is used in order to reject accidentals and recognize missed tones which are masked by larger groups of partials.

For this purpose, different correlated groups of partials are compared in order to recognize repetitive structures in them. The multicorrelation analysis of a chord spectrum described in Chapter 4 provides a means for finding stable groups of partials in a spectrum of a single chord and for rejecting accidental pairwise correlations. Obviously, this approach can be extended to finding repetitive voice spectra in the progression of chords in order to recognize polyphonic parts.

This approach is intensive, since it is based on profound analysis of a given spectral structure and self-organization of data.

Roughly speaking, under the decision making approach we firstly identify two-tone intervals with notes, and then reject some of the notes. On the contrary, under the structural approach we firstly reject some of the intervals, and then identify the remaining ones. Therefore, under the former approach the rejection criterion operates on notes and is based on accounting their frequency, whereas under the latter it operates on the structure of spectral patterns correlated and is based on the simplicity principle.

According to the above classification, we have performed two series of experiments on chord recognition. In the first series based on the recognition of two-tone intervals we have considered chord sequences. For this series of experiments we formulate several important conclusions about the performance of the basic model of interval recognition.

The second series of experiments is based on finding multicorrelated subspectra in spectra of separate chords. For this series of experiments we formulate conclusions about the correspondence of recognition results to certain properties of human perception.

In Section 5.2, "Example of Chord Recognition", we explain the operation of the model with a simple example. We trace the correlation analysis of Boolean spectra of two two-tone chords, according to the model from the previous chapter. We show how accidental patterns can arise and how they can be efficiently recognized in the test experiments. Then we introduce several characteristics with which the performance of the model is studied in the sequel.

In Section 5.3, "Testing the Simple Correlation Approach", we describe the procedures used in computer experiments on interval recognition in detail. In particular, we enumerate and comment on the parameters which are determined for each experiment and the functions of each module used in our test program. Then we explain how the results of a series of experiments are summarized.

In Section 5.4, "Recognition Mistakes", we define two types of recognition errors which occur while recognizing intervals by simple correlation, false notes and missed notes. We analyze the situations which result in these two types of mistakes and suggest several refinements to avoid them, e.g. making spectral representations more accurate. On the other hand, it is explained that refinements in the model are limited by some practical reasons, e.g. the frequency accuracy is limited by the size of time windows. This means that the misrecognition can be reduced but cannot be eliminated completely.

In Section 5.5, "Efficiency and Stability of Recognition", the recognition of two-tone intervals, both harmonic and melodic, is analyzed. Since our model analyzes an unknown number of different pairs of correlated groups of partials and sorts out accidental correlations, the amount of computing and the extent to which the true recognition can be guaranteed are not known beforehand. In order to specify these characteristics, we attempt a special investigation and summarize its results. We conclude that the recognition of chords by recognizing intervals is already quite reliable but cannot be improved within the given simple correlation approach, so that some complements to the model are necessary.

In Section 5.6, "Testing the Decision Making Approach", we consider a model where each note of a chord is determined with taking into account several hypotheses obtained from the analysis of harmonic and different melodic intervals. We illustrate the advantage of the decision making approach and its contribution to the reliability of chord recognition with our series of computer experiments. We also show that the decision making rule may depend on the number of hypotheses processed and that the reliability of final recognition depends on the number of hypotheses as well.

In Section 5.7 "Testing the Structural Approach", we consider a model where the whole interval structure of a chord is recognized simultaneously. We analyze the performance of the model of chord recognition based on multi-correlation analysis of spectral data and on the use of the criterion of least complex data representation. In particular, we show that the limits of correct recognition are similar to the limits of human perception, and that the trends exhibited by the model correspond to that inherent in human recognition.

In Section 5.8, "Judging Computer Experiments", the main conclusions on the performance of the model are formulated.

5.2   Example of Chord Recognition

In order to illustrate the procedure of chord recognition and the way we analyze this procedure, consider two simple chords shown in Fig. 5.1a.

The power spectra of the chords are shown in Fig. 5.1b and their Boolean spectra are shown in Fig. 5.1c. The first chord, its power spectrum, and its Boolean spectrum are precisely the same as in Fig. 4.2, with the only difference that the frequency axis is vertical, in order to make the time axis horizontal.

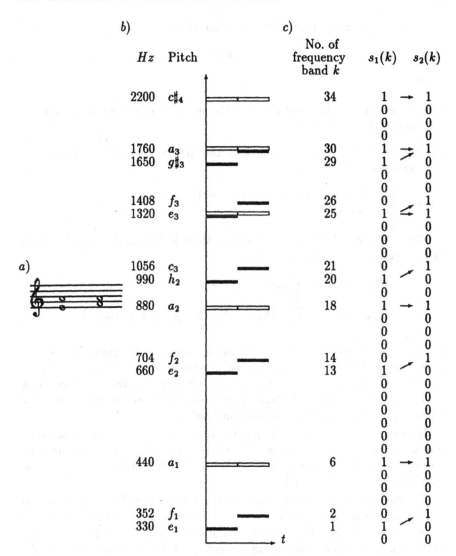

Figure 5.1: Spectral representations of chords

a) chords $(e_1; a_1)$ and $(f_1; a_1)$ in standard notation;

b) their audio spectra (with $\log_2$-scaled frequency axis) for harmonic voices with 5 successive partials (the partials of the lower and upper tones are shown by black or white rectangles, respectively);

c) the Boolean spectra of the chords under the frequency resolution within a semitone — the binary strings $s_n(k)$; the arrows show the parallel motion of partials.

Thus the nth chord is identified with the binary string

$$\{s_n(k)\},$$

where

n is the number of the given chord in the sequence of chords considered (in our example $n = 1, 2$),

k is the numbers of frequency bands in the spectrum considered (in our example $k = 1, \ldots, 34$),

$$s_n(k) = \begin{cases} 1 & \text{if there is a partial in the } k\text{th frequency band,} \\ 0 & \text{otherwise.} \end{cases}$$

We recognize melodic intervals between the nth chord and the $n+1$st chord by the peaks of correlation function

$$R_{n,n+1}(i) = \sum_k s_n(k)s_{n+1}(k+i);$$
$$R_{n,n+1}(-i) = \sum_k s_n(k)s_{n+1}(k-i) = \sum_k s_n(k+i)s_{n+1}(k).$$

If this function has a peak at point i we suppose that the groups of partials correlated can contain tones which constitute the interval of i semitones, while positive i corresponding to ascending intervals and negative i corresponding to descending intervals. The subspectrum correlated is said to be a *generative group of partials* for the given melodic interval.

In other words, a melodic interval between the nth and the $n + 1$st chord is determined by a correlated set of partials in the chord spectrum, and the *salience of melodic interval* of i semitones is identified with the value of the correlation function $R_{n,n+1}(i)$.

The correlation analysis of Boolean spectra of our two chords can be traced in Fig. 5.2. The strings $s_n(k)$ are shown with shifts in order to make the correlation illustrative. The value of correlation function $R(i)$ at point i is equal to the number of 1s coincided in correspondingly shifted strings. For example, $R_{1,2}(-1) = 0$ because there are no 1s coincided in the columns $s_1(k+1)$ and $s_2(k)$.

Table 5.1 displays the most salient melodic intervals not larger than the fifth, i.e. for the values $-7 \leq i \leq 7$. The generative group of partials of each melodic interval is denoted by the indexes of frequency bands. Although we don't use pitch in recognizing intervals, we *conventionally* denote the recognized intervals by notes, referring to the lowest partial of the correlated group of partials.

In Table 5.1 one can see the intervals $(a_1; f_1)$ and $(e_1; a_1)$, corresponding to the voice crossing. If the two voices had different timbre, i.e. different spectra,

```
                No. of
  Pitch       frequency           s₁(k+i)                   s₂(k+i)
              band k

  c♯₄           34        1                         1
                          0 1                       0 1
                          0 · 1                     0 · 1
                          0 · · 1                   0 · · 1
  a₃            30        1 · · · 1                 1 · · · 1
  g♯₃           29        1 1 · · · 1               0 1 · · · 1
                          0 1 1 · · · 1             0 · 1 · · · 1
                          0 · 1 1 · · · 1           0 · 1 · · · 1
  f₃            26        0 · · 1 1 · · ·           1 · · · 1 · · ·
  e₃            25        1 · · · 1 1 · ·           1 1 · · · 1 · ·
                          0 1 · · · 1 1 ·           0 1 1 · · · 1 ·
                          0 · 1 · · · 1 1           0 · 1 1 · · · 1
                          0 · · 1 · · · 1           0 · · 1 1 · · ·
  c₃            21        0 · · · 1 · · ·           1 · · · 1 1 · ·
  h₂            20        1 · · · · 1 · ·           0 1 · · · 1 1 ·
                          0 1 · · · · 1 ·           0 · 1 · · · 1 1
  a₂            18        1 · 1 · · · · 1           1 · · 1 · · · 1
                          0 1 · 1 · · · ·           0 1 · · 1 · · ·
                          0 · 1 · 1 · · ·           0 · 1 · · 1 · ·
                          0 · · 1 · 1 · ·           0 · · 1 · · 1 ·
  f₂            14        0 · · · 1 · 1 ·           1 · · · 1 · · 1
  e₂            13        1 · · · · 1 · 1           0 1 · · · 1 · ·
                          0 1 · · · · 1 ·           0 · 1 · · · 1 ·
                          0 · 1 · · · · 1           0 · · 1 · · · 1
                          0 · · 1 · · · ·           0 · · 1 · · · ·
                          0 · · · 1 · · ·           0 · · · 1 · · ·
                          0 · · · · 1 · ·           0 · · · · 1 · ·
                          0 · · · · · 1 ·           0 · · · · · 1 ·
  a₁             6        1 · · · · · · 1           1 · · · · · · ·
                          0 1 · · · · · ·           0 1 · · · · · ·
                          0 · 1 · · · · ·           0 · 1 · · · · ·
                          0 · · 1 · · · ·           0 · · 1 · · · ·
  f₁             2        0 · · · 1 · · ·           1 · · · 1 · · ·
  e₁             1        1 · · · · 1 · ·           0 1 · · · 1 · ·
                          0 1 · · · · 1 ·           0 · 1 · · · 1 ·
```

	i	0 1 2 3 4 5 6 7	0 1 2 3 4 5 6 7
	$R_{n,n}(i)$	9 1 1 1 2 5 1 3	9 1 1 1 5 2 1 2
	$R_{1,2}(i)$		5 5 0 1 1 5 1 1
	$R_{1,2}(-i)$	5 0 1 1 6 1 1 2	

Figure 5.2: Correlation analysis of chord spectra
(dots denote zeros)

these intervals would be less salient than the intervals corresponding to fluent leading of parts which are constituted by the same voice pattern.

Similarly, we recognize harmonic intervals in the nth chord by the peaks of autocorrelation function

$$R_{n,n}(i) = \sum_k s_n(k)s_n(k+i).$$

The harmonic intervals are determined by correlated groups of partials in the chord spectrum, and the *salience of harmonic interval* of i semitones in the nth chord is identified with the value of the autocorrelation function $R_{n,n}(i)$.

Table 5.2 and Table 5.3 display the most salient harmonic intervals not larger than the fifth, i.e. for the values $0 \leq i \leq 7$, in the first and in the second chord, respectively. Note that the most salient harmonic interval in both tables is the interval of prime ($i = 0$), because the autocorrelation of the unshifted spectrum is always 100%.

The next step in our analysis is detecting the intervals which result from accidental correlations of partials. For example, such a coincidence arises in the descending interval of major third which is specified in Table 5.1 by $i = -4$ with the salience 6. The corresponding generative group of partials contains six partials: Five partials constitute the tone a_1 in the first chord and the tone f_1 in the second chord, and one accidentally joined partial. This accidental is the fifth harmonic of the tone e_1 in the first spectrum ($1650 Hz$) matched to the third harmonic of the tone a_1 in the second spectrum ($1320 Hz$).

In more complex cases, when the chord contains several tones and each tone is constituted by multiple partials, the correlation function may have peaks at the points which do not correspond to real intervals. This can result from accidental correlations of partials which belong to different tones, implying the recognition of these groups of accidentally correlated partials as tones.

As said in Section 5.1, in the first series of experiments we restrict our attention to the performance of the basic model, where harmonic and melodic intervals are recognized by simple correlation. Since we investigate the algorithm, we are interested in characterizing the situations when correlated groups of partials do not correspond to real intervals. In our case of synthesized spectra, such situations are recognized by comparing the correlated groups of partials with the spectral *standard* used in the generation of chord spectra. We characterize a given correlated group by the *number of successive partials* of the standard inherent in the given group.

In our example, the voices have five successive harmonics with the frequency ratio $1 : 2 : 3 : 4 : 5$. According to Example 5 from Section 4.2, the voice standard is determined by the frequency bands indexed by $0, 12, 19, 24, 28$. In Table 5.1 one can find the group of six partials corresponding to the descending interval of major third ($i = -4$) which contains the standard pattern whose partials are indexed by $6, 18, 25, 30, 34$ (that is $0, 12, 19, 24, 28$ translated). Con-

Table 5.1: Most salient melodic intervals between two chords

Correlation $R_{1,2}(i)$	Interval i	Generative group of partials	Number of successive harmonics	Notation of the interval
6	−4	6 18 25 29 30 34	5	$(a_1; f_1)$
5	0	6 18 25 30 34	5	$(a_1; a_1)$
5	1	1 13 20 25 29	5	$(e_1; f_1)$
5	5	1 13 20 25 29	5	$(e_1; a_1)$
2	−7	13 25	2	$(e_2; a_1)$

Table 5.2: Most salient harmonic intervals in the first chord

Autocorrelation $R_{1,1}(i)$	Interval i	Generative group of partials	Number of successive harmonics	Notation of the interval
9	0	1 6 13 18 20 25 29 30 34	5	$(e_1; e_1)$
5	5	1 13 20 25 29	5	$(e_1; a_1)$
3	7	6 13 18	2	$(a_1; e_2)$
2	4	25 30	1	$(e_3; g\sharp_3)$
1	1	29	1	$(g\sharp_3; a_3)$

Table 5.3: Most salient harmonic intervals in the second chord

Autocorrelation $R_{2,2}(i)$	Interval i	Generative group of partials	Number of successive harmonics	Notation of the interval
9	0	2 6 14 18 21 26 30 34	5	$(f_1; f_1)$
5	4	2 14 21 26 30	5	$(f_1; a_1)$
2	5	21 25	2	$(c_3; f_3)$
2	7	14 18	1	$(f_2; c_3)$
1	1	25	1	$(e_3; f_3)$

sequently, the correlated group of partials contains all the five partials of the standard, confirming the right recognition of true interval.

If the number of successive partials was less than five, this would mean that the correlated group of partials doesn't contain the tone pattern used in the chord spectrum generation. In our analysis such a group of partials is immediately recognized as accidental. The insufficient number of successive partials is a simple but efficient criterion for recognizing accidental correlations for testing the performance of the algorithm of interval recognition.

The results of comparing each generative pattern with the standard are displayed in the fourth column of Tables 5.1–5.3. If an interval is rather salient (i.e. if it is characterized by a considerable correlation, meaning that the correlated group of partials is numerous), but the generative group of partials doesn't match the standard, we conclude that we have revealed a *false* interval.

The next step in our analysis is testing the two approaches to reducing the number of mistakes which have been outlined in Section 5.1, decision making approach and structural approach.

The decision making approach is based on the observation that a true tone belongs usually to several harmonic and melodic intervals. If these intervals are identified with certain conventional notes, these notes are recognized several times, whereas the notes corresponding to accidental intervals are most likely recognized only once. Since every chord can be recognized separately by harmonic intervals, or by melodic intervals, being confronted to another chord, the final decision on the chord's compound can be made with taking into account the frequency of the notes in the preliminary hypotheses.

In our model we have considered three hypotheses on each chord from the given progression. The first hypothesis that a given generative group of partials contains a true tone pattern is obtained from the analysis of harmonic intervals of the chord. Another hypothesis is obtained from the analysis of melodic intervals between the given and preceding chord. The third hypotheses is obtained from the analysis of melodic intervals between the given and succeeding chord.

One can consider more hypotheses, analyzing melodic intervals between the given chord and remote chords. For this purpose one should confront the given chord spectrum to spectra of remote chords, as if tracking latent voice leading. However, we restrict our attention to the three hypotheses mentioned, derived from the analysis of the local context.

As shown below, the choice of a decision making procedure depends on the number of hypotheses considered. If we consider all possible hypotheses (as many as there are chords in the analyzed progression), a *majority rule* can be recommended (a tone is accepted if it is backed up by at least a half of the hypotheses). In our case with three hypotheses only, the best results are obtained by their *logical disjunction* (a tone is accepted if it is backed up by at least one hypothesis). c In our simple example, the two decision making rules are equivalent, since there are no mistakes in recognizing tones from salient intervals. However, as seen below, in more complex cases the misrecognition of intervals is quite frequent, and the decision making approach is rather useful.

The structural approach is based on the observation that the structure of partial groups which correspond to real tones is repeated regularly, constituting a stable spectral pattern. On the contrary, accidentally correlated groups of partials which do not correspond to real tones have a random spectral structure. Therefore, in order to recognize an accidental correlation, one has to recognize the irregularity of the structure of the group of partials correlated. This can be done by further correlation analysis applied to different generative groups of partials. If a generative group doesn't correlate with other generative groups, meaning that their structures have little in common, it is almost certain that the given generative group of partials is accidental and then it is rejected. This is the idea of multicorrelation analysis from the previous chapter which filters out accidentally correlated groups of partials. After the whole interval structure of a chord is determined, the groups of partials multicorrelated can be associated with conventional pitches.

The generative groups of partials for our simple example are shown in the third column of Tables 5.1–5.3. One can easily see that all the generative groups of partials with a sufficient salience have the same underlying subspectrum with the partials indexed by $0, 12, 19, 24, 28$. This means that this structure predominates in the generation of the given chords, and after identifying the pitches of its different appearances, one obtains the two chords from our example.

5.3 Testing the Simple Correlation Approach

In order to analyze the procedure of chord recognition, we have performed a series of computer experiments with synthesized spectra of chords.

These spectra have been computed according to certain rules and assumptions by the **first program module**. This module performs the data generation according to some initial parameters. For each experiment the user determines:

1. *Progression of chords* coded in letter names of notes. For example, $(A, e, c\sharp_1, a_1)$ denotes *la major* four-part chord, where capital letters mean the great octave, small letters mean small octave, indexed letters mean the octaves higher than the middle *do*, e.g. a_1 is the *la* of the one-line octave.

2. *Voice type.* We have used *harmonic,* with integer ratio of partial frequencies $1 : 2 : 3 : \ldots$, and *inharmonic,* with the frequency ratio $1 : \sqrt{2} : 2 : \sqrt{2^3} : \ldots$.

3. *Number of partials per voice.* We have used 5, 10, or 16 partials per voice.

4. *Accuracy of spectral representation,* or *frequency resolution.* We have used the accuracy to within 1, 1/2, 1/3, and 1/6 semitone.

5. *Maximal intervals considered.* This parameter is aimed at

   (a) providing means to avoid the octave autocorrelation since odd partials of a harmonic tone can be misrecognized as an independent tone; for this purpose the maximal intervals considered are restricted to 12 semitones;

   (b) restricting the correlation analysis to a limited diapason where the voice spectra are not expected to differ considerably; this effect is known in musical instruments, e.g. the clarinet voice is quite different in low and high registers not only due to the difference in partial intensities but also because of different number of partials.

In our experiments we have used the restriction up to 12 semitones, and no restriction.

For the example from the previous section, the input parameters are written at the top of Table 5.4. Then for each chord its Boolean spectrum is computed in the form of binary string, similarly to that shown in Fig. 5.1c. When the frequency resolution is more accurate than one semitone, the corresponding binary strings become longer, and the 1s become more rare. This results in fewer accidental correlations of partials from different voices.

The **second program module** performs correlation analysis of Boolean spectra. It computes the autocorrelation function for a given chord spectrum, and the correlation function for the spectra of the chord and its predecessor. The result of this procedure is displayed in Table 5.4. Then the program sorts the harmonic and melodic intervals with respect to the correlation values, putting more salient intervals at the top of the corresponding list like in Tables 5.1–5.3 5.4.

The **third program module** processes the most salient intervals, comparing them with the standard voice spectrum which has been used in the spectrum generation. In our example, this standard is determined by the partial frequency ratio $1 : 2 : 3 : 4 : 5$. This comparisons are aimed at estimating efficiency and stability of recognition, and for this purpose certain characteristics of each program run are stored in the memory.

Table 5.4: Correlations of chord spectra

Number of chords $= 2$
Number of partials per voice $= 5$
Ratio of partial frequencies $= 1 : 2 : 3 : 4 : 5$
Accuracy of representation $= 1$ semitone
Maximal intervals considered, up to $= 12$ semitones
Input chords: $(e_1; a_1)$, $(f_1; a_1)$

Number of current chord n	Interval in semi-tones i	Correlation		
		For harmonic interval $R_{n,n}(i)$	For melodic interval from preceding chord	
			Descending $R_{n-1,n}(-i)$	Ascending $R_{n-1,n}(i)$
1	0	9	0	0
	1	1	0	0
	2	1	0	0
	3	0	0	0
	4	2	0	0
	5	5	0	0
	6	0	0	0
	7	3	0	0
	8	0	0	0
	9	2	0	0
	10	1	0	0
	11	1	0	0
2	0	9	5	5
	1	1	0	5
	2	0	1	0
	3	1	1	1
	4	5	6	1
	5	2	1	5
	6	0	1	1
	7	2	2	1
	8	3	2	3
	9	2	2	1
	10	0	0	1
	11	1	3	0

For each chord and each type of interval (harmonic interval, ascending melodic interval, and descending melodic intervals), it is calculated how many intervals from the most correlated ones should be processed in order to reveal all the tones of a given chord. Obviously, the desired information about true harmonic and melodic intervals is represented by the most salient intervals, corresponding to the peak values of correlation function. Therefore, we scan over the intervals, beginning from the most correlated ones.

The minimal set of the most salient intervals of the given type, which contains all true tone patterns of a given chord, is said to be the *sufficient set (of interval patterns)*. The number of these interval patterns conventionally characterizes the *efficiency* of the recognition procedure, corresponding to the amount of necessary computing.

The sufficient set of patterns may contain accidental patterns. The *number of accidental patterns in the sufficient set* also characterizes the efficiency of recognition, corresponding to the amount of "useless" computing. In our simple example there are no accidental patterns, but only accidental partials joined to true patterns. However, in more complex cases salient accidental patterns emerge frequently.

In order to estimate the *stability* of the recognition procedure, we determine the *minimal number of successive standard partials* which may be inherent in true patterns from the sufficient set. If this minimal number is small then all accidental patterns are "very irregular" and, consequently, well distinguishable from true patterns. On the contrary, if this minimal number is close to the number of partials of the standard, then false tones are not well distinguishable from true patterns.

Obviously, the lower is this minimal number of successive standard partials, the larger is the difference between true and accidental patterns, implying the stability of the recognition with respect to possible distortions of data. Certainly, the minimal number of successive standard partials should be regarded with the reference to the number of partials in the standard. For instance, if the standard contains 10 partials then the threshold value 4 is rather small, but if it contains only 5 partials then the same threshold value 4 should be regarded as large, meaning low stability of true recognition.

In order to provide for such a relative comparison, we always display the *maximal number of successive standard partials* inherent in true patterns from the sufficient set.

Note that if the sufficient set is large, the minimal number of successive standard partials is usually large as well (the more patterns, the more the risk of accidentals). From this we conclude that efficiency and stability are interdependent in our model.

The results of such an analysis of the recognition procedure for our simple example is displayed in Table 5.5. In our example, there are no accidental

Table 5.5: Specifications of recognition procedure

Input chords	Sufficient number of patterns/ Number of accidentals in the sufficient set of patterns/ Minimal–maximal number of successive standard partials inherent in true patterns from the sufficient set		
	By harmonic intervals	By melodic intervals	
		From the preceding chord	To the succeeding chord
1. $(e_1; a_1)$	2/0/1–5	—	3/0/1–5
2. $(f_1; a_1)$	2/0/1–5	2/0/1–5	—

partials in the sufficient set. Consequently, the minimal number of successive standard partials is extremely small, being equal to 1, which means high stability. If there was an accidental pattern in the sufficient set, e.g. with 2 successive partials from the spectral standard, then the minimal number of successive standard partials would rise to the value 3.

In a longer progression of chords these estimates are processed as some statistical characteristics and some conclusions on the reliability and stability of the recognition are made for the given run of the program. This is the task for the **fourth program module**.

In a series of experiments where the initial conditions are different, the estimates obtained for each experiment are stored in the memory. Then the results are generalized over all these experiments as well, and general conclusions about the dependence of reliability and stability of recognition on the parameters of the experiments are made. This is the task for the **fifth program module**.

5.4   Recognition Mistakes

The algorithm described commits two types of errors:

- *false (wrong) notes*, and

- *missed (omitted) notes*.

According to such a classification of mistakes, a misplaced note is counted as two mistakes. For example, if the chord (C, E, G) is recognized as (C, D, G), it has one false note D and one missed note E.

Generally speaking, the number of mistakes is the sum of missed and false notes which is the number of notes in the symmetrical difference between

Table 5.6: Specifications of recognition procedure restricted to 12 semitones

Input chords	Sufficient number of patterns / Number of accidentals in the sufficient set of patterns / Minimal–maximal number of successive standard partials inherent in true patterns from the sufficient set		
	By harmonic intervals	By melodic intervals	
		From preceding chord	To succeeding chord
1. $(G; f; c_1; g_1)$*	38/34/11–16 $-g_1$	—	74/59/11–16 $-g_1$
2. $(C; e; c_1; g_1)$*	38/35/11–16 $-g_1$	74/63/11–16 $-g_1$	—

* the chords misrecognized in all of the three hypotheses
— missed tone in the corresponding hypothesis

Table 5.7: Specifications of recognition procedure not restricted

Input chords	Sufficient number of patterns / Number of accidentals in the sufficient set of patterns / Minimal–maximal number of successive standard partials inherent in true patterns from the sufficient set		
	By harmonic intervals	By melodic intervals	
		From preceding chord	To succeeding chord
1. $(G; f; c_1; g_1)$	7/3/11–16	—	16/3/11–16
2. $(C; e; c_1; g_1)$	11/6/11–16	13/6/11–16	—

Figure 5.3: Two hardly recognizable chords

"truth" and "recognition". Thus in the given example

$$
\begin{aligned}
\text{Mistakes} \ &= \ (C, E, G) \bigtriangleup (C, D, G) \\
&= \ [(C, E, G) \cup (C, D, G)] \setminus [(C, D, G) \cap (C, D, G)] \\
&= \ (C, D, E, G) \setminus (C, G) \\
&= \ (D, E).
\end{aligned}
$$

A false note emerges from accidental correlation of partials of irrelevant voices. As mentioned in Section 5.1, accidentally correlated groups of partials have random spectral structure and can be recognized as irregular by comparing them with other generative groups of partials. Being quite efficient, this method nevertheless doesn't guarantee the 100%-reliability.

The case of missed notes is more difficult. In order to illustrate it, consider two four-part chords shown in Fig. 5.3. Suppose that each voice consists of 16 harmonics, and that the accuracy of representation is within 1/3 semitone.

In Tables 5.6 and 5.7 one can see the specifications of two experiments on the recognition of these chords differing in the restriction on the maximal size of intervals considered, up to 12 semitones and with no restriction, respectively.

In the experiment illustrated by Table 5.6 the correct recognition of the two chords is impossible. Consequently, no set of patterns is sufficient. In this case our program scans over all possible intervals, and instead of sufficient number of patterns it prints the total number of patterns scanned (38 for harmonic intervals, and 74 for both types of melodic intervals). Therefore, the number of accidental patterns in the sufficient set of patterns is equal to the total number of accidental patterns (34, or 35 for harmonic intervals, and 63 or 59 for melodic intervals).

While processing an accidental pattern, our program counts the number of successive standard partials in this accidental pattern, which is usually much less than their number in a true tone pattern. In order to find the threshold for distinguishing accidentals, the program determines the minimal number of successive standard partials in true tones which is by one greater than the number of successive standard partials in the accidental patterns processed. In our case this threshold is as high as 11 (with the reference to 16 partials in the voice standard).

Since in the given experiment the harmonic intervals are restricted to 12 semitones, the tone g_1 of the first chord $(G; f; c_1; g_1)$ can be recognized by the only true interval—the fifth $(c_1; g_1)$. Indeed, other intervals which contain g_1 are larger than 12 semitones and cannot be analyzed. However, the correlated groups of partials corresponding to the fifth also contain the partials associated with the interval $(f; c_1)$. This means that the harmonic interval $(f; c_1)$ *masks* the upper parallel interval $(c_1; g_1)$, and the tone g_1 becomes indistinguishable with respect to the only interval of fifth.

In the second chord the note g_1 is also missed when recognized by harmonic intervals. The only true harmonic interval smaller than 12 semitones which contains g_1 is again the interval of fifth $(c_1; g_1)$. However, the correlated groups of partials corresponding to the fifth contain accidentally correlated second overtone of C and first partial of g so that the interval of fifth is misrecognized as $(c; g)$ which "masks" the true interval $(c_1; g_1)$.

The tones g_1 are also missed when the chords are recognized by melodic intervals between them. The only true melodic intervals smaller than 12 semitones which contain g_1 are $(g_1; g_1)$, $(g_1; c_1)$, and $(c_1; g_1)$. However, these intervals are also masked by accidental lower interval $(g; g)$, and true lower parallel intervals $(G; C)$ and $(f; c_1)$, respectively.

In the second experiment illustrated by Table 5.7, with no restriction on the intervals considered, the situation is more favorable. Each tone can be put into correlation with the lowest note of the chord, and thus the corresponding interval cannot be masked by some lower interval, since there are no lower tones. This observation is valid for the recognition by harmonic intervals and by melodic intervals as well.

In spite of the better recognition with no restriction on intervals considered, one should take into account that in reality this restriction is important. It prevents misrecognizing odd partials of a harmonic voice as an independent voice (which results from the octave autocorrelation of harmonic voices). Moreover, the spectra of acoustical sounds change greatly with pitch, which reduces the correlation of tones from different registers even of the same musical instrument. Since low tones are usually rich in harmonics and high tones, on the contrary, quite poor, the audio diapason should be processed in several registers where the tones have similar spectra.

5.5   Efficiency and Stability of Recognition

In order to test our model, we have performed a series of computer experiments on chord recognition by recognizing harmonic and melodic intervals. For the experiments we have chosen the 130th four-part chorale from J.S.Bach's *371 Four-Part Chorales* shown in Fig. 5.4 The chorale has been considered as a sequence of 24 chords, corresponding to the harmonic verticals.

Figure 5.4: The 130th chorale from J.S.Bach's *371 Four-Part Chorales* (the asterisk corresponds to the note missed in recognition)

The Boolean spectra of the chords have been computed and analyzed under various conditions. The experiments have differed in the following parameters listed at the beginning of Section 5.3:

- Voice type (harmonic, with partial frequency ratio $1 : 2 : 3 : \dots$, or inharmonic, with partial frequency ratio $1 : \sqrt{2} : 2 : \sqrt{2^3} : \dots$);

- number of partials per voice (5, 10, or 16);

- accuracy of spectral representation (within 1, 1/2, 1/3, and 1/6 semi-tone);

- restriction on maximal intervals considered (up to 12 semitones, or no restriction).

Having tried all combinations of these parameters, we have performed totally 48 experiments.

A typical computer output for one of these 48 experiments is displayed in Table 5.8. At the bottom of this table one can see some general charac-teristics of the given experiment. The maximal values characterize the worst recognition situations, in particular, when the correct recognition of all true patterns is impossible. The maximal values for recognized chords characterize the amount of computing which is necessary to correctly recognize all recog-nizable true patterns. The average values for recognized chords characterize the average amount of computing for correctly recognizing all recognizable true patterns. Finally, the average values characterize the average amount of computing for processing each chord.

As said in Section 5.3, the results of all the 48 experiments have been summarized by the fifth program module in order to estimate the efficiency and stability of the recognition procedure. The characteristics which are shown

Table 5.8: Specifications of recognition procedure for one experiment on chord recognition
Type of voice: Harmonic
Number of partials per voice: 16
Accuracy of spectral representation: Within 1/2 semitone
Restriction on maximal intervals considered: Up to 12 semitones

Input chords[1]	Sufficient number of patterns / Number of accidentals in the sufficient set of patterns / Minimal–maximal number of successive standard partials inherent in true patterns from the sufficient set		
	By harmonic intervals	By melodic intervals	
		From preceding chord	To succeeding chord
1. $(e; g; e_1; h_1)$	3/0/1–16	—	7/1/9–16
2. $(f\sharp; a; d_1; d_2)$	24/19/10–16 $-d_2$	11/1/4–16	46/33/10–16 $-d_2$
3. $(g; h; d_1; h_1)$	10/5/11–16	8/0/1–16	5/0/1–16
4. $(f\sharp; a; d\sharp_1; h_1)$	8/3/13–16	5/0/1–16	9/1/13–16
5. $(e; g; e_1; h_1)$	3/0/1–16	5/0/1–16	9/1/9–16
6. $(d\sharp; f\sharp; f\sharp_1; h_1)$	3/0/1–16	5/1/10–16	6/2/10–16
7. $(e; g; e_1; c_2)$	4/1/7–16	8/1/7–16	9/0/1–16
8. $(f\sharp; a; d_1; c_2)$	6/2/10–16	5/0/1–16	5/0/1–16
9. $(g; a; d_1; h_1)$	6/1/6–16	5/0/1–16	10/1/4–16
10. $(e; g; d_1; h_1)^*$	24/19/11–16 $-h_1$	46/34/11–16 $-h_1$	46/31/11–16 $-h_1$
11. $(c; g; e_1; a_1)$	24/20/12–16 $-a_1$	8/2/11–16	46/32/11–16 $-a_1$
12. $(A; g; e_1; a_1)$	5/2/11–16	46/36/12–16 $-e_1$	7/2/11–16

13. $(d; f\sharp; d_1; a_1)$	4/0/1-16	9/1/9-16	7/0/1-16
14. $(G; h; d_1; g_1)$	6/3/11-16	10/4/11-16	46/38/11-16 $-G$
15. $(g; g; d_1; h_1)$	5/2/11-16	8/3/11-16	8/1/11-16
16. $(H; g; d_1; d_2)$	24/20/10-16 $-d_2$	5/1/9-16	46/36/10-16 $-d_2$
17. $(d; f\sharp; d_1; a_1)$	4/0/1-16	46/32/9-16 $-a_1$	7/0/1-16
18. $(d; f\sharp; d_1; a_1)$	4/0/1-16	6/0/1-16	46/35/10-16 $-a_1$
19. $(c; a; e_1; e_1)$	2/0/1-16	7/1/9-16	3/0/1-16
20. $(H; h; e_1; g_1)$	3/0/1-16	6/0/1-16	9/1/7-16
21. $(A; h; e_1; g_1)$	5/2/11-16	6/1/11-16	7/0/1-16
22. $(H; h; d\sharp_1; f\sharp_1)$	4/1/8-16	6/0/1-16	7/0/1-16
23. $(H; a; d\sharp_1; f\sharp_1)$	3/0/1-16	8/3/11-16	5/0/1-16
24. $(E; g; h; e_1)$	7/4/11-16	8/2/11-16	46/31/11-16 $-f\sharp_1$
Maximal values	24/20 / 13-16 Fig. 5.5	46/36/12-16	46/38/13-16
Maximal values for recognized chords	10/5 / 13-16 Fig. 5.7	11/4/11-16	10/2/13-16
Average values for recognized chords	5/1 / 6-16 Fig. 5.6	7/1/6-16	7/1/5-16
Average values	8/4 / 7-16	12/5/7-16	19/11/7-16
Number of recognized chords	20	20	16

[1] h denotes note $si\natural$

* the chords misrecognized in three hypotheses

— missed tone in the corresponding hypothesis

in frames are displayed later in diagrams which trace their behavior over all of the 48 experiments.

Now let us outline the main conclusions from our experiments.

The recognition of harmonic voices with 5 partials is independent of the accuracy of representation. This is explained by the precise coincidence of the first 5 harmonics with the degrees of the tempered scale, whence no further separation of partials is possible by refining the spectral representation. This implies that the correlation analysis of spectral data reveals the same intervals under any accuracy of spectral representation.

The harmonic voices with multiple partials are better recognizable when the spectral representation is more accurate. This is explained by the fact that under a poor frequency resolution a large number of partials in voice spectra makes the chord spectrum too dense, resulting in many accidental correlations and thus making difficult distinguishing true patterns from accidentals. Therefore, a refinement of the representation improves the separation of partials of irrelevant voices, implying better recognizability of voices.

For inharmonic voices the recognition is more reliable in case of more accurate frequency representation, both for 5 partials per voice, and for 10 or 16 partials per voice. This is explained by the fact that the partials of inharmonic voices do not fall at the centers of frequency bands which are adjusted to the tempered scale (even in case of 5 partials, unlike harmonic voices). This implies that two partials from two different inharmonic voices, which are indistinguishable under a poor frequency resolution, can be separated in a more accurate representation. That is why accidental correlations of partials of irrelevant voices, which result in accidental interval patterns, occur in an accurate representation more seldom, making chord recognition simpler.

The experiments show that the recognition efficiency and stability depend on voice type, accuracy of representation, and restriction on maximal intervals considered. For the chord recognition by harmonic intervals, these dependencies are displayed by graphs in Fig. 5.5–5.7 where one can trace the behavior of the characteristics shown in frames in Table 5.8 versus representation accuracy and number of partials per voice.

Analogous characteristics for chord recognition by melodic intervals are given in the third and fourth columns in Table 5.8. They have similar properties, so we do not provide figures for them.

The numbers $1, 2, 3, 4, 5, 6$ under the axes of spectral resolution denote the accuracy of representation within $1, 1/2, \ldots, 1/6$ semitone, respectively. (Since in computer experiments we have used only $1, 1/2, 1/3$, and $1/6$, the points corresponding to the accuracy $1/4$ and $1/5$ are linearly extrapolated.) Each diagram displays three cuts, associated with the voices with $5, 10$, or 16 partials. The vertical axis is graduated by 5 units marked by dots.

In Fig. 5.5a and Fig. 5.5c one can see that under the restriction on maximal

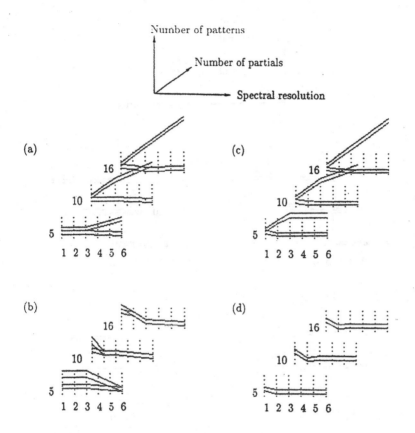

Figure 5.5: Maximal values of the sufficient number of patterns and maximal number of accidentals in the sufficient set of patterns
(upper pair of curves for the whole experiment, the lower pair of chords for only correctly recognized chords in the experiment)

(a) for harmonic voices and maximal intervals considered up to 12 semitones;

(b) for harmonic voices and no restriction on maximal intervals considered;

(c) for inharmonic voices and maximal intervals considered up to 12 semitones;

(d) for inharmonic voices and no restriction on maximal intervals considered.

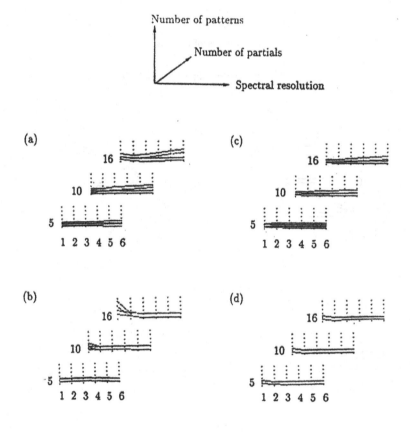

Figure 5.6: Average value of the sufficient number of patterns and average number of accidentals in the sufficient set of patterns
(upper pair of curves for the whole experiment, the lower pair of chords for only correctly recognized chords in the experiment)

(a) for harmonic voices and maximal intervals considered up to 12 semitones;

(b) for harmonic voices and no restriction on maximal intervals considered;

(c) for inharmonic voices and maximal intervals considered up to 12 semitones;

(d) for inharmonic voices and no restriction on maximal intervals considered.

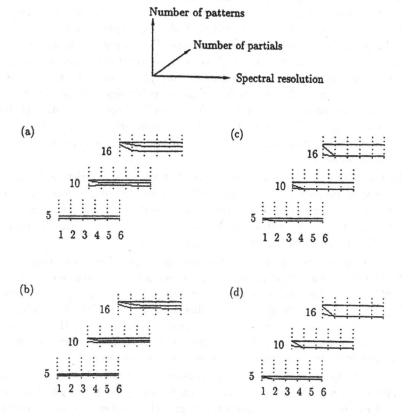

Figure 5.7: Maximal and average values of the limits of the number of successive standard partials inherent in true patterns from the sufficient set
(upper curve is the maximal value of successive standard partials in patterns of sufficient set, the middle curve is the maximum value of minimal number of successive standard partials, the lower curve is the average value of minimal number of successive standard partials)

(a) for harmonic voices and maximal intervals considered up to 12 semitones;

(b) for harmonic voices and no restriction on maximal intervals considered;

(c) for inharmonic voices and maximal intervals considered up to 12 semitones;

(d) for inharmonic voices and no restriction on maximal intervals considered.

intervals considered, the maximal sufficient number of patterns grows with a refinement in the representation accuracy. Indeed, in this case the 100%-correct recognition is impossible, and therefore no set of patterns is sufficient. In our analysis this means that we exhaust the set of possible patterns which is obviously larger for more accurate representations (for example, if the accuracy is within 1 semitone we have 12 intervals per octave, yet if the accuracy is within 1/2 semitone this number is twice greater, implying more interval patterns). The increase in the sufficient number of patterns implies that the number of accidental patterns increases as well. For correctly recognized chords (lower pair of curves in the diagrams), the curves decrease while refining the representation accuracy, because, as mentioned above, salient accidental patterns occur more seldom.

In Fig. 5.5b and Fig. 5.5d one can see that under no restriction on maximal intervals considered the maximal sufficient number of patterns decreases while refining the representation accuracy, i.e. the refinement of representation improves the recognition efficiency. The coincidence of the two pairs of curves means that the 100%-correct recognition is obtained.

Fig. 5.6 illustrates the fact that the average values of sufficient number of patterns and the number of accidental patterns in the sufficient set are much smaller then the maximal values shown in Fig. 5.5. The upper pair of curves in Fig. 5.6a and Fig. 5.6c grows much slower than that in Fig. 5.5a and Fig. 5.5c. This means that the finer is the representation accuracy, the fewer are the cases of misrecognition. In other words, the recognition becomes more reliable, on average requiring fewer patterns to be processed.

The stability of the recognition procedure is illustrated by Fig. 5.7. The distance between the two upper curves can be interpreted as a gap between true and wrong patterns. The larger is this gap, the more seldom an accidental pattern is recognized as a true one. The distance between the two lower curves corresponds to the difference between the greatest similarity of false and true patterns on the one hand, and the average estimation of this similarity on the other hand.

Therefore, the greater the first distance and the smaller the second distance are, the more stable is the recognition procedure. From Fig. 5.7b and Fig. 5.7d one can conclude that the recognition of inharmonic voices is much more stable than that of harmonic voices illustrated by Fig. 5.7a and Fig. 5.7c, and that the recognition stability for inharmonic voices is almost independent of the restriction on maximal intervals considered, which is nevertheless a little better in case when there is no restriction.

5.6 Testing the Decision Making Approach

As mentioned in Section 5.4, the algorithm described commits two types of errors, false notes and missed notes.

False notes emerge under an insufficient accuracy of spectral representation (within a semitone) when accidental correlations are quite probable. For example, when the accuracy of representation is within one semitone, the harmonics after the 12th are so dense in the spectrum (Fig. 3.1) that they fill all successive frequency bands in, implying the correlation of every two groups of high partials.

Therefore, one can make spectral representation more accurate, "rarefying" the spectrum. Then true and false patterns are easier distinguishable, which results in decreasing in the number of false notes in the recognition of chords.

Another cause of false notes is a low salience of voice spectra, when the number of partials per voice is small, which in our experiments corresponds to 5 partials. Indeed, accidentally correlated small groups of partials are more probable than accidentally correlated large groups. Therefore, accidentally correlated groups of partials are better distinguishable from true patterns in case when voice spectral patterns are larger. However, this is true for a sufficiently accurate representation.

On the other hand, refining the accuracy of spectral representation can help in distinguishing true voice patterns with low salience from accidentals. This is caused by the fact that a replication of the same combination, say, of 5 accidental partials is less probable in a rarefied spectrum. For instance, under representation accuracy within one semitone such an accidental correlation requires a coincidence in 5 from 29 frequency bands, yet under accuracy within 1/2 semitone it requires a coincidence in 5 of 57 frequency bands, which is less probable. Such a frequency explanation is most evident for inharmonic voices, since at the standard tempered scale the partials of different inharmonic tones do not coincide with each other and therefore are separable by refining the accuracy of spectral representation.

Missed notes emerge when a true interval pattern is masked by a lower parallel interval (see Section 5.4). Therefore, to avoid missing notes one should avoid lower "maskers". For this purpose the intervals with the lowest tone of a chord can be processed, corresponding to the condition of no restriction on maximal intervals considered.

However, the two ways of eliminating recognition mistakes are efficient only in the model with synthesized data. As already mentioned, the lack of restriction on maximal intervals considered can result in recognizing odd partials of a harmonic voice as an independent pattern. On the other hand, the accuracy of spectral representation of acoustical signal is also limited by the time resolution: The smaller are time windows, the poorer is the frequency accu-

Table 5.9: Preliminary and final results of chord recognition

| Voice type | Maximal number of misrecognized notes in a preliminary hypothesis (over three hypotheses) / Number of misrecognized notes after decision making by a majority rule / Number of misrecognized notes after decision making by logical disjunction of three hypotheses | | | | | |
| | For voices with 5 partials and accuracy of representation | | For voices with 10 partials and accuracy of representation | | For voices with 16 partials and accuracy of representation | |
	1 semitone	1/2 semitone	1 semitone	1/2 semitone	1 semitone	1/2 semitone
Har-monic	8/5/2	8/5/2	10/8/4	7/3/1	18/16/11	7/4/1
Inhar-monic	4/3/1	4/2/0	4/3/1	4/2/0	4/3/1	4/2/1

racy. Moreover, taking into account various disturbances of natural sounds, our method is expected to fail under fine accuracy of spectral representation, since under this condition even small variances in spectra can break the correlation of matched tones.

To a certain extent, these difficulties can be overcome by using a decision making procedure which unites the hypotheses about the chord to be recognized. In our experiments each chord is recognized three times:

- by harmonic intervals within the given chord;

- by melodic intervals between the given chord and its predecessor;

- by melodic intervals between the given chord and its successor.

The only exceptions are the first chord and the last chord, which are recognized in two appearances.

Let us illustrate the effect of such a decision making procedure with an example from our series of experiments. Consider the experiments under the worst conditions, when the accuracy of spectral representation is within 1 or 1/2 semitone and when maximal intervals considered are restricted to 12 semitones.

The most evident way to make the final decision is to apply a *majority rule*, i.e. to accept tones backed up by at least two of the three hypotheses. However, owing to our method of designating notes by the lowest partial of a spectral pattern, missing notes occur much more often than false notes, and better results can be obtained by *logical disjunction of the hypotheses*, i.e. by accepting the tones backed up by at least one hypothesis.

Table 5.9 displays the final recognition results for the 24 chords of the J.S.Bach chorale under these worst conditions. The numbers in the table should be understood in the following way. For example, for the experiment illustrated by Table 5.8 the recognition by harmonic intervals gives 4 mistakes, by melodic intervals from preceding chords—3 mistakes, and by melodic intervals to succeeding chords—7 mistakes. This means that the last preliminary hypotheses gives the maximal number of mistakes 7. After having applied a majority rule to each note recognized, we reduce the number of mistakes to 4 (missed d_2 from the second chord, missed h_1 from the 10th chord, missed a_1 from the 11th chord, and missed d_2 from the 16th chord), and after having applied the logical disjunction, we reduce the number of mistakes is 1 (missed h_1 from the 10th chord, the only mistake inherent in the three hypotheses simultaneously). These indicators for this experiment are displayed in the form 7/4/1 at the right end of the first line of Table 5.9.

One can see that the decision making procedures reduce the number of misrecognized notes. Thus under accuracy of spectral representation within 1/2 semitone the number of mistakes in the worst case is reduced from 8 to

2. Since the four-part chorale represented by 24 harmonic verticals contains $4 \times 24 = 96$, (yet the parts are superimposed two times), 8 mistakes correspond approximately to reliability of recognition 91%, and 2 mistakes to almost 98%. This means that the decision making approach improves the reliability of chord recognition from 91% to 98% (in the worst case).

The recognition reliability can be improved further by considering more hypotheses, derived from correlation analysis of pairs of remote chords. Let us show this with an example of recognizing note h_1 (si_1♮) from the 10th chord in the experiment illustrated by Table 5.8 (marked in Fig. 5.4 and in Table 5.8 by the asterisk).

The failure in recognizing h_1 by harmonic intervals of the 10th chord is caused by the restriction on maximal intervals considered. Indeed, under such a restriction h_1 can be recognized only by the harmonic interval $(d_1; h_1)$ which is masked by the accidental lower parallel pattern $(g; e_1)$ formed by the partials associated with g and the first overtone of e.

The failure in recognizing h_1 by melodic intervals from the previous chord, $(g; a; d_1; h_1)$, is caused by the same reason: h_1 can be detected only from melodic intervals $(h_1; h_1)$, or $(d_1; h_1)$, but the former is masked by the lower true interval $(d_1; d_1)$, while the latter—by the lower accidental pattern $(g; e_1)$.

The failure in recognizing h_1 by melodic intervals form the given chord to its successor, $(c; g; e_1; a_1)$, is also caused by masking melodic intervals $(h_1; a_1)$ and $(h_1; e_1)$, the only ones which can be used in recognition. The former is masked by the parallel accidental pattern $(d_1; c_1)$ determined by the partials associated with d_1 and the first overtone of c. The latter is masked by parallel true interval $(d_1; g)$.

Nevertheless, if we consider broader context and compare the 10th chord with the 12th chord, then the tone h_1 is recognized by melodic interval $(h_1; a_1)$ which is not masked by any pattern of a parallel interval, neither true, nor accidental.

Thus considering the hypotheses derived from confronting remote chords, one can avoid the failure in identifying notes and improve the recognition reliability.

5.7   Testing the Structural Approach

In the previous sections we have investigated the decision making approach to chord recognition. We have shown that if the chords are considered in the context and confronted to a sufficient number of other chords, they can be correctly recognized by recognizing two-tone intervals. In this section we discuss the structural approach to chord recognition which is based on simultaneously finding all the intervals of a chord, in other words, on finding the multi-interval

Figure 5.8: Chords used for testing the recognition algorithm

Table 5.10: Recognition of chord $C_7 = (c; e; g; b)$

Accuracy in	Number of partials per voice			
semitones	5	10	16	32
1	Correct	Correct	$(c; e; g; b; c_1)$	$(c; e; g; b; c_1)$
1/2	Correct	Correct	Correct	$(c; c_1; e_1; g_1; b_1; c_2; e\flat_2)$
1/3	Correct	Correct	Correct	Correct
1/4	Correct	Correct	Correct	Correct

structure of a chord.

Since recognizing intervals is the basic procedure in recognizing acoustical contours and trajectories, the conclusions concerning the efficiency and stability of recognition by simple correlation analysis are also important for the recognition by multicorrelation analysis of chord spectra. In particular, the estimates of sufficient number of patterns is important for limiting the number of the generative groups of partials (tone patterns) to be processed at the triple-correlation stage of the algorithm from Section 4.4.

The model of chord recognition described in Section 4.4 has been tested on the second series of experiments. Recall that this model is based on multicorrelation analysis of a chord spectrum and uses the criterion of least complexity of representation of spectral data in order to choose the optimal representation of the chord structure.

In our tests we have used 2 four-note chords and 3 five-note chords shown in Fig. 5.8 (we use b for $si\flat$ and h for $si\natural$). For each chord we have synthesized its Boolean spectrum, having used harmonic voices with 5, 10, 16, or 32 successive partials under the accuracy of spectral representation within 1, 1/2, 1/3, and 1/4 semitones. Having tested 5 chords in 16 different conditions, we have performed totally 80 experiments.

The recognition results for each of the five chords are displayed in Tables 5.10–5.14. Let us make some remarks concerning the recognition results referring to Table 5.13.

As seen from tracing the rows of the Table 5.13 from left to right, adding more partials to the voice pattern makes the recognition more difficult under

Table 5.11: Recognition of chord $C_{7M} = (c; e; g; h)$

Accuracy in semitones	Number of partials per voice			
	5	10	16	32
1	Correct	$(c; ab; c_1; eb_1; g_1)$	$(c; c_1; e_1; g_1; h_1; c_2; eb_2; e_2; f_2)$	$(c; ab; c_1; eb_1)$
1/2	Correct	Correct	$(c; e; g; h; c_1)$	$(c; ab; c_1; eb_1)$
1/3	Correct	Correct	Correct	Correct
1/4	Correct	Correct	Correct	Correct

Table 5.12: Recognition of chord $C_{7/9} = (c; e; g; b; d_1)$

Accuracy in semitones	Number of partials per voice			
	5	10	16	32
1	Correct	$(c; e; c_1; d_1; e_1)$	$(c; e; g; b; c_1; d_1; e_1; g_1)$	$(c; e; g; b; c_1; d_1; e_1)$
1/2	Correct	$(c; e; g; d_1)$	Correct	$(c; c_1; d_1; e_1; g_1)$
1/3	Correct	$(c; e; g; d_1)$	Correct	Correct
1/4	Correct	$(c; e; g; d_1)$	Correct	Correct

Table 5.13: Recognition of chord $C_{7M/9} = (c; e; g; h; d_1)$

Accuracy in semitones	Number of partials per voice			
	5	10	16	32
1	Correct	$(c; e; g)$	$(c; g; ab; b; c_1; eb_1)$	$(c; g; ab; b; c_1; eb_1)$
1/2	Correct	Correct	Correct	$(c; c_1; d_1; e_1; g_1; ab_1; a_1; h_1; c_2)$
1/3	Correct	Correct	Correct	$(c; e; g; h; c_1; d_1; e_1; g_1)$
1/4	Correct	Correct	Correct	Correct

Table 5.14: Recognition of chord $C_{6/9} = (c; e; g; a; d_1)$

Accuracy in semitones	Number of partials per voice			
	5	10	16	32
1	Correct	Correct	$(c; e; g; a; c_1; d_1; e_1; g_1)$	$(c; e; g; a; c_1; d_1; e_1)$
1/2	Correct	Correct	Correct	$(c; a; c_1; d_1; e_1)$
1/3	Correct	Correct	Correct	Correct
1/4	Correct	Correct	Correct	Correct

the same accuracy of representation. In case of representation accuracy within 1 semitone, the recognition is right for the voices with 5 partials. For voices with 10 partials, only the *chord type* is recognized, i.e. the fundamental major triad $C = (c; e; g)$ instead of the full harmony $C_{7M/9} = (c; e; g; h; d_1)$. For voices with 16 partials there emerge notes which are incompatible with the original harmony. These notes are enhanced by bold face style: **ab**, **b**, and **eb₁**. One can see that neither the chord, nor the chord type are correctly recognized. The harmony recognized is $Ab_{7M/9}$ instead of original harmony $C_{7M/9}$.

As seen from tracing the columns of Table 5.13 from top to bottom, any refinement of the representation accuracy improves the recognition. In case of 32 partials per voice, the recognition is totally wrong under the representation accuracy within 1 semitone. Under the representation accuracy within 1/2 semitone, there are less notes incompatible with the original harmony. Next, under the accuracy within 1/3 semitone, the chord type is already recognizable. Finally, under the accuracy 1/4 semitone, the recognition is completely correct.

In order to compare these trends with that in human perception, we have performed audio tests with synthesized chords. These tests confirm the same trends in audio perception: The chord perceptibility is better for voices with fewer partials, and for better sharpness of hearing (we interpret the accuracy of frequency representation as the sharpness of hearing). Besides, the audio experiments show that the chords generated by tones with 32 and even 16 partials of equal power are hardly recognizable being perceived as noise. Similar difficulties in the computer recognition are seen in Tables 5.10–5.14.

Our experiments show that the chord (its contour) can be recognizable even if the generative tone spectrum is recognized with mistakes. The misrecognition of generative tones is caused by the reduction of generative patterns in our model while finding the least complex representation of spectral data (see Section 4.4). For example, translations of a generative pattern may result in coincidences of some partials from different voices, and the generative pattern can be reduced without any change in the chord spectrum (see Fig. 4.3). Such a representation is less complex than that with the original generative pattern, and therefore it is preferable. However, the pitch of such a spectral pattern can be hardly identifiable (i.e. the chord tones are not recognizable).

The recognizability of chords independently of the recognizability of tones illustrates our earlier remark, proving that high-level patterns can be recognizable even if the low-level patterns are not recognizable (see Fig. 2.2 from Section 2.2). Similarly, the type of a chord (major or minor) is recognizable regardless of the recognizability of whole chords; in our experiments we have seen that the type of chord is recognized easier than the full harmony.

Thus our model exhibits the following successive degrees in chord recognition: The chord type (major or minor) is recognized first, next goes the recognition of harmony (the chord to within permutation of notes) then the recognition of chord with its interval structure, and the chord tones are recognized worst. This precisely corresponds to human perception.

5.8 Judging Computer Experiments

Let us enumerate principal conclusions of the chapter.

1. We have realized a series of experiments on the recognition of chords with a computer model of correlative perception. At first we have investigated the basic (simple correlation) model for recognizing two-tone intervals in chords and chord progressions. Harmonic intervals of a chord are recognized by peaks of the autocorrelation function of the chord Boolean spectrum with a $\log_2$-scaled frequency axis. Melodic intervals between two chords are recognized by peaks of the correlation function of the chord Boolean spectra. The recognition of intervals is performed with no reference to absolute pitch and with no previous learning. The model is applicable to the recognition of harmonic and inharmonic voices.

2. The recognition of two-tone intervals by correlation analysis of chord spectra is tested on its efficiency and stability. It is shown that the method is more efficient and stable when the accuracy of spectral representation increases, and the intervals considered are not restricted to a certain limit. However, it is mentioned that such conditions are not realistic in practical applications. To overcome this difficulty, some special measures are proposed in order to improve the performance of the model.

3. We have proposed two approaches to improve the basic (simple correlation) model of chord recognition, decision making approach and structural approach. The final recognition by decision making approach is performed with regard to the frequency of certain tones in the totality of two-tone intervals found. The recognition by structural approach is based on finding two-tone intervals which are generated by voice patterns with the same spectral structure. It is shown that both approaches considerably improve the reliability of chord recognition.

4. The decision making approach has been tested on recognizing four-part Bach polyphony for synthesized spectra under various conditions, both for harmonic and inharmonic voices. Totally, 48 experiments have been performed. Owing to the use of the decision making option, the recognition reliability in our 48 experiments has been improved, e.g. under the accuracy of spectral representation within 1/2 semitone the recognition reliability in worst cases was improved from 93% to 98%.

5. The structural approach has been tested on recognizing four-part and five part separate chords for synthesized spectra under various conditions. Totally, 80 experiments have be performed. It has been shown that the recognition capacity of the model is close to the limits of human perception. Moreover, the trends in the performance of the model are the same as in human perception: The chord recognizability is better for voices with fewer partials and for more accurate spectral representation, interpreted as better sharpness of musical hearing. The chord type (major or minor) is recognized best, next goes the recognition of harmony (chord to within permutation of notes), then the recognition of the chord interval structure, and tones (and their pitch) are recognized worst.

Chapter 6

Applications to Rhythm Recognition

6.1 Problem of Rhythm Recognition

The nature of time, rhythm, and tempo is one of the first questions posed by music theory. Many distinguished music theorists have contributed to the understanding of related perception mechanisms. Recently, a series of works has been published where the problem of rhythm and tempo is formulated with regard to computer rhythm recognition and tempo tracking (see Section 1.2 for references).

However, further progress in modeling rhythm and tempo perception is constrained by the lack of explicit definitions of these concepts. The known definitions are rather ambiguous. For example, rhythm is defined as the order and the proportion of durations (Porte 1977); tempo is explained as a characteristic of execution motion with respect to measures and melodic, harmonic, rhythmic, or dynamical cues (Pistone 1977); time is considered as a form of determination of rational proportions of rhythm (Viret 1977). The incompleteness of these definitions is obvious; however, according to Dumesnil (1979), one can hardly find better definitions, even in special psychomusicological publications like (Fraisse 1983; Povel & Essens 1985).

The definitions of tempo and rhythm are not only incomplete but also interdependent. On the one hand, the tempo is defined with respect to a certain rhythm (if there are no events, no motion can be perceived). On the other hand, in order to measure the proportion of durations, the rhythm is defined with respect to a certain tempo. This makes a kind of logical circle.

Attempting to overcome such an ambiguity in defining interdependent concepts, we apply the principle of correlativity of perception. According to this principle, time data are represented in terms of repetitious low-level configurations and a high-level configuration of their relationships. The grouping of time events into low-level configurations is realized with respect to the simplicity principle, meaning that the representation of time events by high-level and low-level configurations requires least memory.

In our case, the low-level configurations are correlated *rhythmic patterns.* The high-level configuration, being determined by time relationships between the rhythmic patterns correlated, is associated with the *tempo curve.* In other words, we consider repetitious rhythmic patterns as recognizable reference units for tempo tracking. Drawing analogy to vision, similar rhythmic patterns correspond to instant states of an object, and the tempo curve corresponds to the object trajectory in time. Table 6.1 shows the analogy between rhythmic patterns and visual patterns in Fig. 2.1. Note that the tempo curve is not supposed to be continuous, which meets Desain & Honing's (1991) view at tempo phenomenon.

As mentioned in Section 2.2, the advantage of such a representation is the recognizability of high-level patterns regardless of the recognizability of low-level patterns. This means that we may have difficulties in identifying a rhythm as waltz, march, etc., while being capable to recognize its tempo. On the other hand, if the tempo has been determined, the rhythm recognition becomes easier.

The goal of this chapter is developing a technique for finding rhythmic patterns. First of all we consider the problem of rhythm recognition in general, and then restrict our attention to the segmentation of rhythmic progressions with respect to timing accentuation. For this purpose some formal rules of timing accentuation and classification of rhythmic patterns are introduced. Then the rules of segmentation are illustrated with an example of rhythm recognition.

In Section 6.2, "Rhythm and Correlative Perception", we formulate the problem of rhythm recognition from the standpoint of the principle of correlativity of perception. In case of rhythm, the two-level scheme of data representation is generalized to a multi-level scheme, since the rhythm is represented as a hierarchy of embedded rhythms. Such a multiple embedding makes the rhythm redundant and thus easier recognizable under considerable tempo fluctuations.

In Section 6.3, "Correlativity and Recognition of Periodicity", we apply the principle of correlativity of perception to the recognition of quasi-periodicity in a sequence of time events, corresponding to the recognition of repetitions under variable tempo. In particular, we trace the operation of the method of variable resolution with a simple example and compare the results obtained by our model with some conclusions from known psychological experiments.

In Section 6.4, "Timing Accentuation", we pose the problem of rhythmic segmentation. In order to solve this problem we formulate rules for distinguishing certain events which are called accentuated. In the given study we consider

Table 6.1: Correspondence between visual and time data

	Visual data	Time data
Stimuli	Pixels	Time events
Low-level patterns	Symbols A	Rhythmic patterns
High-level pattern	Symbol B	Tempo curve

the accentuation with respect to time cues only, ignoring pitch and dynamic cues. We define strong and weak accents and illustrate these definitions with an example.

In Section 6.5, "Rhythmic Segmentation", we show that the accentuation alone is not sufficient for unambiguous segmentation of time events. In order to realize the segmentation and classification of segments, the notion of rhythmic syllable is introduced. A rhythmic syllable is understood to be a sequence of time events with the only accent at the last event. Referring to a simple psychoacoustic experiment, we show that rhythmic syllables are perceived as indecomposable rhythmic units.

In Section 6.6, "Operations on Rhythmic Patterns", we consider a kind of rhythmic grammar. We define the elaboration of a rhythmic pattern as a subdivision of its durations which preserves the original pulse train. The elaboration is also explained in terms of correlation of rhythmic patterns. The junction of rhythmic syllables is defined to be the elaboration of their sum. This way we define possible transformations of rhythmic patterns other than distortions caused by tempo fluctuations. Using the concepts of elaboration and junction we explain certain properties of rhythm organization.

In Section 6.7, "Definition of Time and Rhythm Complexity", we define a root pattern to be the simplest pre-image (with respect to the elaboration) of a generative syllable of the rhythm. In order to recognize such a pattern the above mentioned concepts of rhythmic syllable and their junction are used. The notion of root pattern is applied to time determination and estimation of rhythm complexity.

In Section 6.8, "Example of Analysis", we consider the snare drum part from Ravel's *Bolero* and trace the operation of our model of rhythm segmentation step by step. As a result, the sequence of time events is represented in terms of generative syllables and their transformations. This representation reveals root patterns, their rhythmic elaboration, and thus the structure of the rhythm. Finally, the representation obtained is used to determine the time of the rhythm.

In Section 6.9, "Summary of Rhythm Perception Modeling", the main statements of the chapter are recapitulated.

6.2   Rhythm and Correlative Perception

As mentioned in the previous section, our approach to rhythm and tempo recognition is based on a two-level representation of time events in terms of generative rhythmic patterns and their transformations. The transformations fall into

- distortions which are caused by tempo fluctuations, and

- variations which are caused by musical elaboration.

We shall consider both types of transformations, using special techniques for each type.

Firstly, restrict our attention to distortions caused by tempo fluctuations, supposing that there are no elaboration of rhythmic patterns. The problem is to interpret a sequence of time events either in terms of rhythm, or in terms of tempo, or both. For example, consider the sequence of time events shown in Fig. 2.3. In Fig. 2.4a this sequence of events is interpreted in terms of rhythm, i.e. as a single rhythmic pattern under a constant tempo. In Fig. 2.4b this sequence of events is interpreted in terms of tempo, e.g. as a repeat of the first three durations (generative rhythmic pattern) under a tempo change. Moreover, every sequence of time events can be interpreted as generated by a fixed single duration which is getting shorter or longer because of tempo changes at each event. On the other hand, it can be interpreted as a single complex rhythmic pattern under a constant tempo (cf. Fig. 2.3–2.4 from Section 2.2). However, the two extreme representations are complex, the former owing to the complexity of the tempo curve, and the latter owing to the complexity of the rhythmic pattern. We suppose that in most cases there exists a compromise representation, with a few rather simple generative rhythmic patterns and a rather simple tempo curve, while the total complexity being quite moderate.

The choice of interpretation can be influenced by tradition, context, or some special intentions. In Section 2.2 we have shown that the interpretation may depend on melodic context. For this purpose we have estimated the complexity of alternative representations. In case of a pure rhythm, the complexity is least for the representation of time events as a single pattern, whereas in case of melodic context its representation as a repeat is preferable.

This example illustrates the idea that the recognition of rhythmic patterns determines the perception of tempo. If in Fig. 2.4a and Fig. 2.4c the three quavers are recognized as a repet of the three crotchets, then the tempo is perceived as changing. If the group of the last three durations is not recognized as a repeat of the first three durations, then the tempo is perceived as constant.

At the same time, the recognition of rhythmic patterns depends on tempo. Indeed, rhythmic patterns are perceived as repetitious, only if they are comparable in time, i.e. if they are considered with respect to a certain tempo.

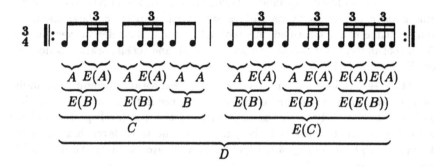

Figure 6.1: Multilevel repetition structure of rhythm from *Bolero* by M.Ravel

Therefore, rhythm and tempo are interdependent concepts, which implies the impossibility of their separate recognition. Nevertheless, in the model of correlative perception their functions are different, and the interaction between them is determined with respect to the criterion of least complex representation.

Thus the task of rhythm and tempo recognition is formulated as finding the least complex representation of a sequence of time events. This representation reveals the regularity of time organization which in case of Western music is usually observed at several levels simultaneously (Lerdahl & Jackendoff 1983), corresponding to simultaneous perception of musical texture at several levels, e.g. at the level of measures, couples of measures, etc., up to constituents of musical form.

Therefore, a model of rhythm perception should be multi-level, while each level being characterized by its own generative patterns determined by generative patterns of lower level. These levels should be commensurable with respect to tempo, otherwise each would have its own tempo curve, which would complicate the representation of data. The desired representation of a given sequence of time events can be understood as a tree with indecomposable rhythmic segments at the low level, and branches which show their grouping into the patterns of higher levels.

In order to recognize the multi-level regularity, we have to consider the second type of transformations, variations of rhythmic patterns caused by musical elaboration. Let us illustrate the idea of multi-level regularity with an example of snare drum part from Ravel's *Bolero* (Fig. 6.1).

At the first level the regularity emerges as the commensurability of the quaver durations which are braced in the first subscript line in Fig. 6.1. Since quaver durations are repeated regularly, the pulse train of quavers is quite evident.

According to Mont-Reynaud & Goldstein (1985), a subdivision of a rhythmic pattern is said to be its *elaboration*. With respect to elaboration, the quaver patterns are classified into *root patterns* and their derivatives. In Fig. 6.1 a root pattern is denoted by a letter, e.g. A, and its elaboration is denoted by $E(\cdot)$, e.g. $E(A)$.

At the second level, the rhythm regularity emerges as the commensurability of crotchet durations and their elaborations, braced in the second subscript line in Fig. 6.1. Similarly to rhythmic patterns of the first level, patterns of the second level are grouped into the patterns of the third level, braced in the third subscript line in Fig. 6.1, which in turn are grouped into the patterns of the fourth level.

The idea of such an arrangement of rhythmic patterns is the representation of the *embedded pulse train of the rhythm*. Thus the rhythm of *Bolero* is characterized by the following four embedded levels:

1. Quavers (elaborations of root pattern A).

2. Crotchets (elaborations of root pattern B).

3. Three-quarter measures (elaborations of root pattern C).

4. Repetitious two-measure segments D.

Such a multilevel representation of rhythm is used further for classification of rhythmic patterns and determination of time.

The embedded pulse train is inherent only in so called *divisible rhythms* which are used in Western music. The rhythmic structure of other cultural traditions is not described by such multi-level schemes.

It is remarkable that the tempo is most free in Western music as well. This can be explained by the fact that several embedded pulse trains make a rhythm redundant (since there are several cues to recognize repetitions) and thus recognizable even under severe distortions. Similarly to the example in Fig. 2.4c–d, where a repetitious rhythm is complemented with a repetitious intonation, a redundant rhythm is complemented with an embedded rhythm of another level, implying its total representation as a repeat to be less complex than its representation as a single rhythmic pattern. Hence, a redundant rhythmic structure is recognizable even under considerable tempo deviations.

The range of tempo fluctuations which do not prevent from perceiving repetitions is larger if there are several complementary cues like melodic intonation, similarity of accompaniment, harmonic pulsation, etc. For example in Skrjabin's performance of his *Poem* Op. 32 No. 1 transcribed from a piano-roll recording (Skrjabin 1960) the tempo varies within limits $\downarrow = 19 \div 110$, i.e. up to 5.5 times. These fluctuations are unambiguously perceived as tempo changes but not as changes of durations (changes of rhythmic patterns), owing to several complementary cues in the repetitious texture like accompaniment figures, phrasing, etc.

Tempo changes are strictly prohibited in Bulgarian or Turkish music with so called *additive rhythms* with complex duration ratios. From the standpoint of the correlativity principle it is explained by the fact that if a rhythm is not structurally redundant, then even minor tempo deviations are not perceived as *accellerando* or *ritardando* but rather as changes of durations (rhythmic changes), implying an inadequate perception of musical meaning.

Thus we have shown that the rhythm recognizability under tempo deviations depends on the rhythm redundancy caused by rhythmic elaboration. This dependence means that in a model of rhythm recognition the two types of transformations of rhythmic patterns, caused by tempo fluctuations and musical elaboration, should be taken into account simultaneously.

6.3 Correlativity and Recognition of Periodicity

In the remainder of the chapter we discuss different elements of our approach to rhythm recognition, detection of periodicity, accentuation, segmentation, and time determination.

Further we use the notions of *periodicity* and *rhythm*. By periodicity we mean a repetition of events or their groups with no segmentation into rhythmic patterns. If time events are periodically segmented with respect to the time structure, we obtain the concept of rhythm.

At first, consider the problem of recognizing periodicity. In our model of correlative perception the recognition of repetitious rhythmic patterns is based on autocorrelation analysis of a sequence of time events under various distortions of time scale. The distortions which provide high correlation of patterns can be found by the method of variable resolution introduced in Section 2.4.

Let us trace the operation of the model of correlative perception with a simple example of recognizing quasi-periodicity (generative element) in a sequence of time events in Fig. 6.2.

Consider the following sequence of tone onsets recordered to within one time unit which is equal to 1/20 sec. (Fig. 6.2a):

$$t = 0, 10, 19, 30.$$

Define the *event function*, for all $t = 0, 1, \ldots, 30$ putting

$$s(t) = \begin{cases} 1 & \text{if } t = 0, 10, 19, 30, \\ 0 & \text{otherwise.} \end{cases}$$

The event function $s(t)$ in a form of binary string is shown in Fig. 6.2b. One can try to determine its period p by finding the peaks of autocorrelation function

$$R(p) = \sum_t s(t - p)\, s(t).$$

However, no autocorrelation and, consequently, no periodicity is recognizable (see the second column in Table 6.2). According to the method of variable resolution, the accuracy of representation in Fig. 6.2b should be reduced as shown in Fig. 6.2c. Then autocorrelation function $R(p)$ has salient peaks (see the third column in Table 6.2) and the periodicity is recognizable. After the correlated events have already been known, these events should be locally shifted in order to increase in the correlation. The result of this procedure is shown in Fig. 6.2d. Restoring the resolution, we obtain the sequence in Fig. 6.2e together with the description of tempo fluctuations corresponding to the shifts of time events determined earlier.

Now one has to accept or reject the hypothesis about the representability of this sequence of time events as generated by a repetitious rhythmic pattern, which in the given case is a single duration of 10 units. For that purpose one has to compare the complexity of two alternative representations. The first representation corresponds to storing the period of 10 units with a repeat algorithm (or its call) provided with a tempo curve. The second representation corresponds to coding the given sequence of time events as it is, i.e. as a single rhythmic pattern under a constant tempo. Obviously, the complexity of total representation depends on the way of coding repeat algorithm and tempo curve.

By some reasons, usual algorithms of coding functions (by means of storing the coefficients of their polynomial approximations or Fourier series) are not suitable for our purposes. Usual ways of coding require the knowledge of the function values on the whole domain of its definition, whereas we would like to track the tempo in real time, coding the tempo curve while processing current data.

One of possible ways of coding the tempo curve is fixing the time moments when the tempo deviates from its current value, say, by more than 5%. Such a heuristic algorithm meets zonal properties of perception and its logarithmic sensitivity to absolute values. On the other hand, it is quite simple, which is desirable in computer experiments.

The model of recognition of rhythm periodicity has been tested with a series of computer experiments on the recognition of rhythm from *Bolero* (Fig. 6.1) performed by striking a computer keyboard. The experiments have revealed the repetitious structure of the rhythm at four levels shown in Fig. 6.1. The repetition has been recognized also at the level of sixteenth triplets, even in cases when their duration ratio 1:1:1 has been distorted up to the ratio 7:6:10.

Such a great variance of relative durations observed for sixteenth triplets

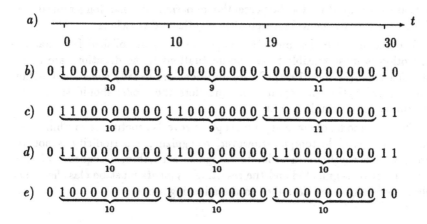

Figure 6.2: Representation of time events with variable resolution

Table 6.2: Autocorrelation $R(p)$ of time events

p	$R(p)$ in Fig. 6.2b	$R(p)$ in Fig. 6.2c	$R(p)$ in Fig. 6.2d	$R(p)$ in Fig. 6.2e
...	0	0	0	0
8	0	1	0	0
9	1	3	3	0
10	1	4	6	3
11	1	3	3	0
12	0	1	0	0
...	0	0	0	0
18	0	1	0	0
19	1	3	2	0
20	1	3	4	2
21	0	1	2	0
22	0	0	0	0
...	0	0	0	0
29	0	1	1	0
30	1	2	2	1
31	0	1	1	0

hasn't been observed for longer durations. This means that these short durations are less important for the perception of periodicity than longer durations, which in our experiments correspond to eighths and crotchets.

The above conclusion meets the experimentally established fact that the perception is most sensible to tempo fluctuations if the durations are about 0.1–1.0 second (Michon 1964). These durations are usually considered as fundamental in rhythm perception, and therefore the model should treat these durations with a certain priority.

Thus the model of correlative perception reveals repetitious rhythmic patterns under variable tempo. However, recognizing a periodicity is not yet recognizing a rhythm. In order to recognize a rhythm, a sequence of time events must be segmented and the resulting segments must be classified. The related procedures are described in the following sections.

6.4   Timing Accentuation

As mentioned at the beginning of the previous section, in order to be considered as a rhythm, a periodical sequence of time events should be segmented, i.e. certain events must be recognized as starting points of periods.

A distinct periodicity can lack an unambiguous segmentation, as in African percussion music where almost every time event may be considered as a start point of a period (Schloss 1985). In such a case, illustrated by Fig. 6.3 we say that we can recognize no rhythm but just a periodicity.

Thus in order to perceive a rhythm unambiguously, one has to recognize both periodicity and start points of periods which are accentuated somehow. This means that we have to distinguish between accentuated and non-accentuated events. All the known ways of accentuation are based on breaking the homogeneity in pitch, harmony, dynamics, timbre, etc.

In the present study we restrict our attention to timing accentuation. In speech the timing accentuation is realized by pauses and longer vowels, i.e. by longer durations. Generally speaking, the same is valid for music.

Summing up what has been said, let us introduce the following rules of timing accentuation which are similar to some rules formulated by Boroda (1985; 1988; 1991).

Rule 1 (Durations) *The only characteristic of a time event is the duration of time interval between its onset and the onset of the next time event. This interonset time interval is said to be the duration associated with the event. If the given time event is the last in the sequence, the associated duration is assumed to be not fixed.*

This rule restricts the attention to timing cues only. An important remark

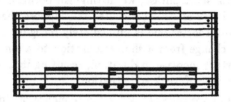

Figure 6.3: Ambiguous segmentation of a periodical sequence of time events

concerns the last time event whose duration can be arbitrarily long. The use of this assumption will be clear in the sequel.

Rule 2 (Accentuation Distinguishability) *In order to distinguish accentuated events in a sequence of time events, at least two types of durations are necessary.*

This rule postulates the case when timing accentuation is possible.

Rule 3 (Accentuated Durations) *The duration associated with an event is said to be strongly accentuated if*

(a) it is longer than its closest neighbors, i.e. it follows a shorter duration and the next one is also shorter;

(b) it follows an equal duration and the next one is shorter.

The duration is said to be weakly accentuated if

(c) it follows a shorter duration and precedes an equal duration which is not strongly accentuated, i.e. the second next is not shorter.

A time event is said to be accentuated (strongly or weakly) if the duration associated with the event is accentuated (strongly or weakly, respectively).

The idea of the third rule is that a longer duration which is adjacent to a shorter one is accentuated. The most evident case (a) concerns a situation when a longer duration is between two shorter ones. If a duration is between shorter and equal one then it is usually accentuated (case *b* and *c*), yet in order to avoid simultaneous accents at two equal successive durations between two shorter ones, we assume no accent at the first duration (case *c*). No accentuation emerges when durations are successively getting shorter or longer.

In order to provide unambiguous segmentation when several accentuated durations emerge in a short phrase, we distinguish between strong and weak accents, with the priority of strong accents. We suppose that a change from a longer duration to a shorter one is immediately recognized, resulting in an accent. Yet after a change from a shorter duration to a longer duration one can expect some further increase in duration, resulting in a weaker sensation of accent; this is the case of weak accentuation.

To illustrate the above rules, consider the sequence of time events shown in Fig. 6.4. The first crotchet marked by symbol ">" is weakly accentuated by virtue of Rule 3c, since it is the duration between a shorter duration and an equal one. The last crotchet which is also marked by ">" is strongly accentuated by virtue of Rule 3b, since it is the duration between an equal and a shorter one. Note that although the second crotchet precedes an eighth, it is not accentuated. Indeed, by virtue of Rule 1 the duration of the event associated with this eighth is crotchet, and no shorter duration is adjacent to it. By virtue of Rule 1 the last note of the sequence can be considered both as accentuated, or not.

To show the accentuation in notation, bar lines are put before accentuated events as shown in Fig. 6.4. In the above example, the segmentation with respect to the accentuation determines the 3/4 time of the given phrase.

6.5   Rhythmic Segmentation

The accentuation defined is not sufficient for rhythmic segmentation. Indeed, consider a periodic sequence of time events segmented with respect to the accentuation defined in two different ways as shown in Fig. 6.5a and Fig. 6.5b. To prove the perceptual ambiguity of segmentation of these events, we have performed the following audio experiment: The given sequence of time events has been recorded and played back in a loop, having been amplified gradually from zero level. A series of audio tests has shown that listeners recognize the two segmentations with almost equal probability.

Thus to recognize a rhythmic segmentation we need some other cues in addition to the accentuation. For that purpose we introduce rules of classification and elaboration of rhythmic patterns. By a *rhythmic pattern* we understand any segment of a given sequence of durations. However, the most important is the case when rhythmic patterns are segmented with respect to the accentuation. Thus we obtain the following definition.

Rule 4 (Phrases and Syllables) *A rhythmic phrase is defined to be a sequence of durations which follows an accentuated duration and ends at an accentuated duration. A rhythmic phrase with the only accentuated duration is said to be a rhythmic syllable (Katuar, 1926).*

Figure 6.4: Accentuation by timing cues

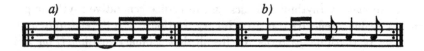

Figure 6.5: Rhythmic segmentation by timing cues

Consequently, a rhythmic syllable is a simplest rhythmic phrase. Any rhythmic phrase is formed by adding syllables to each other.

Note that by virtue of Rule 4 a syllable is determined by the durations which precede an accentuated event. The accentuated duration itself is not included into the syllable. The accentuated event just marks the end of the syllable, and the associated accentuated duration may be not fixed (cf. with Rule 1).

We suppose that rhythmic syllables are perceived as indecomposable time units. In order to prove it we have performed the following audio experiment: The rhythmic syllable shown in Fig. 6.6 with two fixed absolute durations $0.1sec$ has been repeatedly reproduced under variable delays divisible by $0.1sec$, e.g. 0.8, 1.0, 1.2, 0.8, ... sec. If the rhythmic syllable was perceived as a composed structure, the common time unit ($0.1sec$) would result in a sensation of constant tempo with changes of rest durations. (Tempo determination with respect to the common time unit was proposed by Messiaen (1944)). However, in our audio experiment listeners have recognized tempo deviations rather than rhythmic changes. This proves that the syllable is perceived as an entirety rather than as composed of smaller units and that the end duration is not important for identifying equal syllables.

Such a way of recognizing tempo by time intervals between the entries of

Figure 6.6: Syllable as an indecomposable unit

similar rhythmic patterns meets the principle of correlativity of perception. In fact, in our experiment we have shown that similar rhythmic patterns are used as reference indivisible units for tempo tracking. Besides, we have shown that the tempo is a percept of another level than the rhythm.

6.6   Operations on Rhythmic Patterns

In order to classify rhythmic phrases and recognize generative rhythmic patterns, we define a reflexive transitive binary relation E, "*is the elaboration of*" on the set of rhythmic patterns X. Recall that a binary relation E on X is reflexive if xEx for all $x \in X$ (a rhythmic pattern is the elaboration of itself), and transitive if xEy and yEz implies xEz for all $x, y, z \in X$ (a successive elaboration of a rhythmic pattern is its elaboration).

Rule 5 (Elaboration) *Rhythmic pattern A is the elaboration of rhythmic pattern B if A preserves the pulse train of B, i.e. if A results from a subdivision of durations of B by inserting additional time events (Mont-Reynaud & Goldstein, 1985).*

Fig. 6.7 illustrates the idea of rhythmic elaboration with an example of subdivisions of a crotchet duration (recall that by virtue of Rule 1 the crotchet duration, in order to be determined, should be followed by a next tone onset which is not shown in the figure).

The idea of elaboration can be explained in correlation terms. Represent the rhythmic patterns in Fig. 6.7 by 0 and 1 within the accuracy of a sixteenth. Then the top pattern which we denote by T is written down as follows

$$T = \{t_1 \ldots t_4\} = \{1000\},$$

and the bottom pattern which we denote by B is written down as

$$B = \{b_1 \ldots b_4\} = \{1111\}.$$

It is easy to see that pattern B is the elaboration of pattern T if and only if $B \supset T$. This means that B contains all ones of T. Since the number of ones in T is equal to the autocorrelation

$$R_{T,T} = \sum_{i=1}^{4} t_i \cdot t_i$$

(which in the given case is equal to 1), and the number of coinciding ones in B and T equals to the correlation

$$R_{B,T} = \sum_{i=1}^{4} b_i \cdot t_i$$

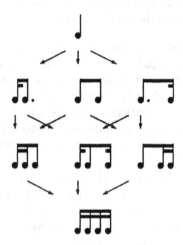

Figure 6.7: The elaboration of a crotchet rhythmic pattern

(which in the given case os equal to 1), we obtain that B is the elaboration of T if and only if

$$R_{B,T} = R_{T,T}.$$

Since the correlation is usually understood as a measure of similarity, the last equation means that the pattern B, being the elaboration of pattern T, is similar to pattern T.

For the patterns of equal duration which are not the elaboration of each other (as in the second line of Fig. 6.7), the correlation is less than autocorrelation. For example, putting

$$L = \{l_1 \ldots l_4\} = \{1100\}$$

and

$$M = \{m_1 \ldots m_4\} = \{1010\},$$

we obtain

$$R_{L,M} = \sum_{i=1}^{4} l_i \cdot m_i = 1 < 2 = R_{L,L} = \sum_{i=1}^{4} l_i \cdot l_i = R_{M,M} = \sum_{i=1}^{4} m_i \cdot m_i.$$

Now we define the junction of syllables.

Rule 6 (Sum and Junction of Syllables) *The sum of two successive rhythmic patterns is defined to be the rhythmic pattern constituted by the time events of these patterns which are put one after another.*

The junction of two successive rhythmic syllables is defined to be a rhythmic syllable which is the elaboration of their sum.

Note that the sum of two syllables is more than the two syllables in succession. Besides the two syllables themselves, the sum contains the *link*—the accentuated duration after the first syllable. According to the remark following Rule 4, this duration is undefined if the first syllable is considered separately, since instead of the whole duration we consider just an accent. In the sum of syllables, this accent turns to be a duration, linking the two syllables. Therefore, there can be many different sums of the same two syllables, depending on the link duration.

Also note that the sum of two rhythmic syllables is a rhythmic phrase, whereas their junction is a rhythmic syllable. This means that the sum of two syllables can have two accents, at the ends of each syllable, whereas in their junction the internal accent is suppressed by dividing the associated duration into shorter ones which are no longer accentuated. This implies that the junction has a new rhythmic quality, a through tension towards its end.

Fig. 6.8a displays two identical rhythmic syllables and their junction. The total duration of the third syllable is the same as the sum of the two syllables. This results in the symmetry of the whole passage, providing its structure to be $1 + 1 + 2$.

Consider another junction of the two syllables, for instance, obtained by adding two quavers to the third syllable as shown in Fig. 6.8b. This implies that we link the first two syllables not by a crotchet duration but by a half-note duration, i.e. we consider the elaboration of another sum of the syllables.

The connection between the three syllables in Fig. 6.8b is less evident than in Fig. 6.8a. Indeed, in Fig. 6.8a one can see not only the two syllables, but already their sum which is elaborated next. In a sense, the elaboration is already "prepared" for easy perception. On the contrary, in Fig. 6.8b the sum of the two syllables is different from the sum which is elaborated. In Fig. 6.8b the intermediate phase between the two syllables and their junction is missed, breaking the successiveness in their perception.

We could provide the effect of such a successiveness in Fig. 6.8b, making the rest between the first two syllables longer, up to a half-note duration. The duration of the second rest is not so important. Even if we change the duration of the second rest in Fig. 6.8a, the third rhythmic syllable is still perceived as the elaboration of the sum of the first two.

From our standpoint, we can explain the simplicity of rhythmic construction $1 + 1 + 2 + 4 + \ldots$. Such a structure contains a rhythmic pattern, the elaboration of the preceding segment, then the elaboration of two preceding segments, and so on. Therefore, the origin of such a structure is quite simple, adding junctions of all preceding segments. This results in perceiving such rhythms with ease; moreover, the perception is "prepared" to recognize the elaboration since the sum is already exhibited.

Figure 6.8: Two different junctions of the same rhythmic syllables

6.7 Definition of Time and Rhythm Complexity

Thus we have introduced the rules of representation of a given sequence of time events in terms of generative syllables. Constructing such representations, one can reveal origins of a given rhythm with conclusions concerning its time.

Note that rhythmic patterns of equal total duration constitute an *ordered directed set* with respect to the elaboration, where every two elements have a common superior—their common *root*. An example of such an order is shown in Fig. 6.7 with a common root pattern at the top and its successive elaborations indicated by arrows.

The patterns of the same total duration which are not elaborations of each other (like in the second line of Fig. 6.7) are of particular interest. If a rhythm contains such patterns then this rhythm has no embedded levels of the pulse train and can be represented as a succession of irreducible units whose pulse train becomes predominant.

The idea of a pulse train generated by indecomposable rhythmic patterns can be applied to rhythmic syllables. Since each syllable has the only accent, the accents of syllables determine a pulse train with a certain rhythm. We use this rhythm to determine the time of a given sequence of time events.

Rule 7 (Determination of Time) *If a sequence of time events is representable in terms of elaboration of certain rhythmic syllables (phrases), then the time of the given sequence is determined by the duration ratio of their roots.*

In other words, one has to find a stable preimage (with respect to the elaboration) of generative patterns.

Roughly speaking, the time is defined to be the rhythm of roots of generative syllables. In a sense, time patterns, being superior to generative patterns in the hierarchical representation, constitute an intermediate representation level between rhythmic patterns and the tempo curve (cf. Fig. 6.1).

Besides time determination, rhythmic patterns which are irreducible to each other can be used for estimating the complexity of rhythm. Indeed, their number corresponds to the number of generative patterns required to generate the given sequence.

For example, consider the rhythm in Fig. 6.9 which is constituted by two rhythmic patterns of equal duration. One can see that the crotchet duration is the root for the two rhythmic groups beamed but no rhythmic group is the elaboration of another. This means that the pulse train of crotchets is supported by no pulse train of quavers or some other shorter durations.

Such a rhythm can be considered as less redundant and therefore as more complex. The *complexity of a rhythm* can be identified with the branching index of the graph of the rhythmic patterns used, i.e. by the maximal number of irreducible to each other rhythmic configurations of the same level.

As seen from Fig. 6.1, each rhythmic level of *Bolero* is generated by a single pattern, implying its complexity being equal to 1. The rhythm in Fig. 6.9 is generated by two patterns of equal duration which are not reducible to each other; consequently, its complexity is equal to 2.

Such an understanding of rhythm complexity meets the ideas of Messiaen (1944) who has characterized the diversity of rhythm by the number of non-commensurable patterns used.

Thus finding irreducible (with respect to elaboration) patterns has two applications: time recognition and estimation of rhythm complexity.

6.8   Example of Analysis

Consider the snare drum part from *Bolero* by M.Ravel (Fig. 6.10). Since we use time data only (Rule 1), our method cannot be applied to the rhythms which are based on pitch and dynamic accentuation. Since the chosen rhythm contains two types of durations, by virtue of Rule 2 it is an appropriate object for our analysis. Let us trace the procedure of structurizing this rhythm step by step.

1. Consider Duration 0. The following one is shorter, consequently, by virtue of Rule 3b it is strongly accentuated. Since it is the first event in the sequence, we recognize the first syllable S as constituted by Duration 0 only. To write down the syllables, we shall use the denotations from Section 6.6, with the only difference that a digit will correspond not to

Figure 6.9: A rhythm with the indicator of complexity 2

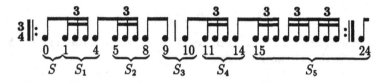

Figure 6.10: Determination of time by recognizing rhythmic syllables

the duration of sixteenth but to the duration of sixteenth triplet. Thus,

$$S = \{100\}, \text{ corresponding to } \text{♩.}$$

Thus up to the current moment our rhythm is represented by the only syllable

$$S.$$

2. Consider Duration 1. It is preceded by a longer duration and succeeded by an equal one. By Rule 3 it is not accentuated. By Rule 4 we don't recognize the end of a syllable at Duration 1.

   Since Durations 2 and 3 are not preceded or succeeded by shorter ones, by virtue of Rule 3 they are not accentuated. Since they are not accentuated, by Rule 4 we don't recognize the end of syllable at these durations.

3. Since Duration 4 is between two shorter durations, by virtue of Rule 3a it is strongly accentuated. By Rule 4 we recognize the end of syllable which we denote

$$S_1 = \{111\ 100\}, \text{ corresponding to } .$$

Now we compare syllable S_1 with the earlier recognized, verifying:

(a) whether the given syllable is the elaboration of another one;

(b) whether any other syllable is the elaboration of the given one;

(c) whether the given syllable is the junction of other syllables;

(d) whether any other syllable is the junction of the given syllable with another one.

One can see that syllable S_1 is not the elaboration of any other syllable, no syllable is the elaboration of S_1, but S_1 is the junction of two syllables S. Therefore, up to the current moment our rhythm is represented as

$$S \quad S_1$$

or

$$S \quad E(2S),$$

where $E(2S) = E(S + S)$ denotes the elaboration of the sum $S + S$ (i.e. the junction of two syllables S).

4. Similarly to Item 2, there is no accentuation at Durations 5–7 and we don't recognize the end of syllable.

5. Similarly to Duration 4 analyzed in Item 3, there is an accentuation at Duration 8, with the only difference that Duration 8 is *weakly* accentuated. By Rule 4 we recognize the end of syllable which we denote

$$S_2 = \{111\ 100\}, \text{ corresponding to } \text{♩♩♩♪}.$$

Note that syllable S_2 is equal to S_1. Consequently, everything said about syllable S_1 relates also to S_2. Therefore, up to the current moment our rhythm can be represented in the following two ways

$$\begin{array}{ccc} S & E(2S) & E(2S); \\ S & S_1 & S_1. \end{array}$$

6. One can see that Duration 9 is not accentuated, and therefore no syllable ends at Duration 9.

7. By Rule 3b Duration 10 is accentuated, and we recognize syllable

$$S_3 = \{100\ 100\}, \text{ corresponding to } \text{♩♪ ♩♪}.$$

Answering the questions (a)–(d) enumerated in Item 3, we recognize that S_1 and S_2 are the elaborations of S_3; besides, S_3 is the sum of two syllables S. Thus we obtain the following equivalent representations of our rhythm:

$$\begin{array}{cccc} S & E(2S) & E(2S) & 2S; \\ S & E(S_3) & E(S_3) & S_3. \end{array}$$

8. Since Durations 11–13 are not accentuated, no syllable ends at these durations.

9. Since by Rule 3a Duration 14 is accentuated, we recognize syllable

$$S_4 = \{111\ 100\}, \text{ corresponding to } \text{♩♩♩♪}.$$

Having answered the questions (a)–(d) enumerated in Item 3, we obtain the following representations of the rhythm:

$$S \quad E(2S) \quad E(2S) \quad 2S \quad E(2S);$$
$$S \quad E(S_3) \quad E(S_3) \quad S_3 \quad E(S_3).$$

10. Since Durations 15–23 are not accentuated, no syllable ends at these durations.

11. By virtue of Rule 3a Duration 24 (or Duration 0, taking into account the repeat sign) is accentuated. Consequently, we recognize syllable

$$S_5 = \{111\ 111\ 111\}, \text{ corresponding to } \text{♩♩♩♩♩♩♩♩♩ | ♩}.$$

Having answered the questions (a)–(d) enumerated in Item 3, we obtain that

$$S_5 = E(S_1 + S_3) = E(S_2 + S_3) = E(S_3 + S_3).$$

Hence, we get the following two representations of our rhythm:

$$S \ \|: \ E(2S) \quad E(2S) \quad 2S \quad E(2S) \quad E(4S) \ :\| \ ;$$

$$S \ \|: \ E(S_3) \quad E(S_3) \quad S_3 \quad E(S_3) \quad E(2S_3) \ :\| \ ,$$

or

$$S \ \|: \ S_1 \quad S_1 \quad S_3 \quad S_1 \quad E(S_1 + S_3) \ :\| \ . \tag{6.1}$$

If we consider strong accents only, ignoring weak accents, then syllables S_2 and S_3 join into syllable

$$S_{2+3} = \{111\ 100\ 100\ 100\}, \text{ corresponding to } \text{♩♩♩♩ ♩♩ | ♩}.$$

Since rhythmic syllable S_5 is the junction of syllables S_2 and S_3, we obtain even more simple representation of the rhythm as follows

$$S \ \|: \ S_1 \quad S_{2+3} \quad S_1 \quad E(S_{2+3}) \ :\| \ . \tag{6.2}$$

With regard to the repetitions of the given rhythm, syllable S can be interpreted as the end of syllable S_5. Finally, we obtain the representation of the given rhythm as generated by phrase S_1, S_{2+3}. Since S_{2+3} is two times longer than S_1, by virtue of Rule 7 we interpret our rhythm as having triple time: 3/4, or 3/8, etc. The choice of denominator (unit of counting) is a question of convention.

Note that there is a risk to interpret the period in (6.1) as consisting of three equal groups, i.e. instead of "correct" segmentation

$$S \; \|: \; [S_1 \quad S_1 \quad S_3] \quad [S_1 \quad E(S_1 + S_3)] \; :\| \; ,$$

on can accept the "wrong" segmentation

$$S \; \|: \; [S_1 \quad S_1] \quad [S_3 \quad S_1] \quad [E(S_1 + S_3)] \; :\| \; .$$

This corresponds to recognizing the time of the rhythm as 2/4. However, the representation (6.2) which is obtained by ignoring local accents leaves no doubts in the triple time basis. Thus distinguishing between strong and weak accents is rather useful.

6.9   Summary of Rhythm Perception Modeling

Summing up what has been said, let us enumerate the main items of the chapter.

1. The proposed approach to rhythm recognition is based on some general principles (correlativity of perception, optimal data representation), on some heuristic methods (the way of coding the tempo curve and estimating the complexity of rhythm), and on some particular properties of hearing (priority of certain durations in rhythm and tempo perception).

2. The interaction of rhythm and tempo is understood as follows. Rhythmic patterns are considered as reference time units for tempo tracking. The interdependence of rhythm and tempo is overcome by the least complex data representation, implying the juxtaposition of rhythm and tempo in an "optimal" way. Since the model is destined for recognizing any forms of repetitions, it is applicable both to divisible and additive rhythms.

3. In order to realize a directional search for generative rhythmic patterns, we suggest formal rules of accentuation and segmentation. Accents are associated with longer durations which determine a segmentation of time events into rhythmic syllables. Then elaboration, sum, and junction of rhythmic syllables are defined. Using this kind of rhythmic grammar, we represent a series of time events in terms of generative syllables (phrases) and their transformations. The method is illustrated with an example of time determination of a given rhythm.

Chapter 7

Applications to Music Theory

7.1 Perception Correlativity and Music Theory

In the present chapter we discuss the correspondence between our model and music theory. This is important, because, on the one hand, music theory generalizes the experience of auditory perception acquired during centuries and, on the other hand, these generalizations are not influenced by any *a priori* assumptions which on the contrary are always inherent in scientific experiments.

Recall that basic assumptions of the given work have been formulated under the influence of the author's studies in music theory (see Section 1.4). In this chapter we compare the initial statements with the results obtained. One can expect that such a logical circle will not introduce anything principally new. However, refinement of the principle of correlativity of perception, its mathematical formulation, and computer modeling makes it possible to comment on music theory with an advanced understanding of the subject and complete initial observations with new conclusions.

The material of this chapter falls into two parts: Applications of our model to psychoacoustics (perception scales, relative and absolute hearing) and music theory in a proper sense (harmony, counterpoint, orchestration). Some earlier mentioned applications are discussed here as well.

In Section 7.2, "Logarithmic Scale in Pitch Perception", we adduce reasons in favor of logarithmic scale in pitch perception and insensitivity of the ear to the phase of the signal. It is supposed that these two properties are necessary for the decomposition of complex sounds into constituents, thus contributing to recognizing the causality in sound generation. In particular, owing to these two properties of hearing, musical tones are perceived as entire sound objects rather than as collections of sinusoidal partials, yet chords are perceived as built of complex tones.

In Section 7.3, "Definition of Musical Interval", we generalize the definition of musical interval to sounds with no pitch salience (as bell-like sounds). The interval is defined to be the distance of translation of a tone spectral pattern in the frequency domain. To measure this distance, the tones have not to be harmonic, and no coordinates of the tone location (fundamental frequencies) are needed. Thus the pitch is eliminated from the definition of interval. This meets the known fact that most people are capable to recognize intervals but cannot recognize pitch. Finally, we mention that timbral changes can influence on the perception of intervals and chords.

In Section 7.4, "Function of Relative Hearing", we explain that relative (interval) hearing, i.e. the capacity to perceive intervals juxtaposed to absolute hearing which is the capacity to recognize pitch, is the correlative perception in the frequency domain. Drawing analogy to vision, we suppose that interval hearing is the capacity to estimate the translation distance between similar sound objects without determining their precise location. This implies that interval hearing is based on the recognition of structure by identity and structure by similarity in audio scenes, contributing to separation and tracking simultaneous acoustical processes.

In Section 7.5, "Counterpoint and Orchestration", we discuss the difference between certain rules of counterpoint and orchestration. In particular, the use of parallel voice-leading in orchestration, which is prohibited in counterpoint, is justified as a mean for creating sonorous effects. Then we comment on such orchestration rules as using a common note for transmitting a melody from one instrument to another, using bright timbres for solo parts and soft voices for orchestral "pedals", precaution against large spacing, etc.

In Section 7.6, "Harmony", some properties of classical harmony are regarded as providing conditions for the recognizability of chords in our model, which is interpreted as providing conditions for adequate music perception. We explain a certain acoustical advantage of the scale of just intonation over the equal temperament. Then we comment on the prohibition against voice-crossing and voice-overlapping, prescription to fill in skips in voice-leading, function of ornamentation, role of bass in harmony, incompleteness of cluster chords, best recognizability of triads in the root position, and interaction between music hearing and music memory.

In Section 7.7, "Rhythm, Tempo, and Time", the three concepts enumerated are defined from the standpoint of the principle of correlativity of perception. We suppose that they characterize three different levels of correlative perception of time events. We also explain why minor changes of time data can considerably change the perception of rhythm and tempo. This follows from the instability of optima with respect to data changes.

Finally, in Section 7.8, "Summary of the Chapter", we recapitulate the main statements of the present discussion.

7.2 Logarithmic Scale in Pitch Perception

As follows from Chapter 3, the logarithmic scale in pitch perception together with the insensitivity of the ear to the phase of the signal predetermine some remarkable properties of perception.

Owing to logarithmic scaling, patterns with a linear structure, like tone spectra with harmonic ratio of partial frequencies $1 : 2 : \ldots : K$, being non-linearly compressed, become irreducible. This corresponds to the known fact that harmonic tones are perceived as entire sound objects rather than as compound ones.

The irreducibility of harmonic spectra is proved for power spectra, i.e. not for usual sound spectra with complex coefficients, but for spectra with real positive coefficients. This means that harmonic tones can be perceived as entire sound objects if only the ear is insensitive to the phase of the signal. Indeed, by virtue of Lemma 1 from Chapter 3, discrete spectra with complex coefficients are isomorphic to polynomials over complex numbers. By the fundamental theorem of algebra, such polynomials are always factored into polynomials of the first degree. This means that a spectrum with complex coefficients can be always decomposed into the convolution product of spectra with complex coefficients constituted by two close impulses each. In case of sound spectra, this property would imply total decomposability of all sounds (see Sections 3.5 and 3.7).

Therefore, the insensitivity of the ear to the phase of the signal is a useful property rather than an imperfectness of hearing. Together with the logarithmic scaling of pitch, it prevents from a "trivial" decomposition of sound which has no physical sense. These two properties of hearing are necessary for the decomposition of sound corresponding to different physical sources, thus contributing to the recognition of causality in the acoustical environment.

The indecomposability of harmonic tones is an important prerequisite for their use in music. Besides, harmonic tones are acoustically compatible with speech and with each other, resulting in consonant harmonies. Moreover, they are distinguishable in chords and in polyphony, and their timbral diversity is sufficiently large. Another important property of harmonic tones is the fact that high-level patterns of their relationships (chords and melodic lines) are quite stable with respect to noise, distortions, and voice changes. All of this make harmonic tones to be universal basic elements of musical compositions and a reliable carrier of semantical musical information.

Thus one can conclude that the logarithmic scale in pitch perception and the insensitivity of the ear to the phase of the signal not only contribute to the recognition of causality in sound, but also predetermine the use of tones with pitch salience as basic musical elements.

7.3   Definition of Musical Interval

Recall that an interval between two musical tones is usually defined as the difference between the tone pitches, or, which is equivalent, as the ratio of their fundamental frequencies (Gelfand 1981; Gut 1976). Similarly, chords and melodies are classified according to ratios of fundamental frequencies of their tones (Frances 1972). In other words, the known definitions of interval, chord, and melodic line are based on the idea of pitch.

Such definitions suppose that the perception of pitch precedes the perception of intervals. However, most people perceive intervals but fail in note identification, distinguish between major and minor chords not recognizing their roots, and sing songs transposing them arbitrarily instead of singing them in a fixed key.

These observations show that the mechanism of interval hearing is independent of the pitch recognition capability. Moreover, since most listeners recognize intervals but rather few recognize pitch, it is the interval hearing which is predominant in music perception. In this connection we formulate a definition of interval, chord, and melodic line not using the concept of pitch. Since chords and melodic lines are constituted by intervals, the definition of interval is principal, and definitions of chord and melodic line are its derivatives.

To illustrate our approach to defining intervals between tones with no reference to pitch, we draw analogy to the distance between visual objects. Let us show that in certain cases the distance between objects can be defined, even if it is difficult to define their precise location.

For example, consider a square on the plane shown in Fig. 7.1. The definition of its coordinates is not evident: They can be identified either with its bottom left angle, or with its center, or with some other point. In any case, the definition of coordinates of a complex object requires an additional convention. However, there is no other way of defining coordinates of a complex object other than by representing the object by a certain point.

The distance between objects is usually defined as the distance between their representative points, like mean points, centers of gravity (Fig. 7.2), or some others chosen by special rules, like in case of the Hausdorff distance between two sets. However, if the objects are *equal* and *equally oriented* in space then the distance can be defined with no reference to representative points but directly, by the magnitude of corresponding translation of the objects. This is illustrated in Fig. 7.3 where the idea of distance between two equal squares is quite evident. Since each point of the square is translated by the same distance d, all the points are equally representative, and there is no need to single out one of them in order to represent the square. The distance is measured between "entire" objects, but not between their representative points whose

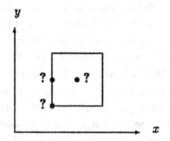

Figure 7.1: Ambiguity in defining coordinates of a complex object

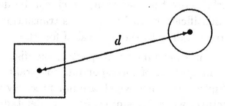

Figure 7.2: Distance between dissimilar objects

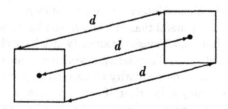

Figure 7.3: Distance between similar objects

coordinates must be determined.

Thus *if objects are equal and equally oriented in space then the distance between the objects is measured directly, without attributing the distance to some representative points of the objects.*

The need for attributing the distance to some points emerges if only the objects are dissimilar, or differently oriented in space like in Fig. 7.2. However, it should be taken into account that a reference to special points complicates the idea of distance, requiring for an additional information. According to the simplicity principle in perception, any complication should be avoided, and the estimation of distance between equal objects does provide such a possibility.

The above approach to defining the distance between equal objects with no reference to representative points can be applied to defining intervals between musical tones. Since the pitch is a kind of representative point of sound objects, the distance between similar tones can be estimated with no reference to their pitch. For example, consider two tones shown in Fig. 3.2. Since the two spectra are equal and equally oriented in "acoustical space" (due to the unidimensionality of the frequency domain), the distance between them can be measured by analogy with Fig. 7.3. For this purpose it suffices to measure the distance of translation of the spectrum, which can be done directly, independently of pitch identification. In Fig. 3.2 this translation distance is equal to five semitons, corresponding to the interval of fourth.

Therefore, if two tones have identical spectra, the distance between them can be defined as the magnitude of corresponding spectral translation. We can also extend this definition to tones which are not precisely identical but similar in spectral structure, where by similarity we understand high correlation between tone spectra. Spectra of two similar tones can be regarded as generated by the same spectral pattern which is translated along the $\log_2$-scaled frequency axis with slight distortions of its envelope (linear filtering). The translation distance is said to be the *interval* between the tones.

Although our definition of interval is restricted to *tones with similar spectral structure*, it is more general than the known one based on the distance between tone pitches. Indeed, all musical sounds (which have clear pitch salience) satisfy the condition of spectral similarity, having harmonic ratio of partial frequencies $1 : 2 : \ldots : K$. The similarity of harmonic spectra is quite evident at a logarithmic frequency axis. Even if spectral envelopes are different, the structure of partial frequencies of harmonic tones is invariant with respect to pitch translations, providing for a high correlation of the tone spectra. Therefore, two harmonic spectra can be considered as resulting from a translation of a spectrum with envelope distortions. Consequently, the definition of interval proposed is applicable to all *harmonic* tones (with a pitch salience). This means that our definition covers the case considered in the traditional definition.

Besides, our definition of interval is applicable to *inharmonic* (bell-like) sounds with no pitch salience, or even band-pass noises with similar spectra. For example, the sound of two bells has the same spectral structure, whence the translation distance between the corresponding similar spectra can be measured without pitch identification.

In a sense, a harmonic tone with unambiguous pitch is analogous to a visual object with the only evident representative point, as in case of circle with its center. An inharmonic sound with ambiguous pitch is analogous to a visual object with no evident representative point, as in case of square in Fig. 7.1. Using the pitch for the determination of intervals is analogous to measuring the distance exclusively between objects whose representative points are uniquely determined, implying that we can measure the distance between circles but not squares. This evident absurdity shows the imperfectness of traditional definition of interval.

Our definition of interval, being independent of the shape of sound objects (harmonic or inharmonic tones), is analogous to the fact that the distance in Fig. 7.3 is independent of the shape of visual objects (circles or squares). This analogy proves that our definition of interval is quite natural.

If sound spectra differ considerably, as in case of violin voice and bell sound, they cannot be regarded as generated by translations of the same spectral pattern, and our direct definition of interval is not applicable. In this case the choice of representative points is necessary in order to determine the distance between sound objects (cf. the distance between dissimilar objects illustrated in Fig. 7.2). However, since the pitch of inharmonic sounds is ambiguous, the traditional definition of interval in terms of pitch is not applicable either.

Our definition of interval implies the dependence of interval recognition on voice timbres. Imagine the following experiment: At first, an interval between two similar (inharmonic) tones is determined by a salient peak of correlation function of their spectra. Then the spectral envelopes are changed gradually, in order to amplify certain partials and suppress some others, as in experiments of Shepard (1964) and Risset (1971; 1978). Gradually transforming (filtering) the spectra in this way, one can suppress the initial peak of the correlation function and amplify another peak. This means that the initial interval becomes ambiguous. One can continue this transformation of voice spectra, making the new peak predominant, implying that the new interval becomes more salient than the former. This way timbre changes can result in interval transformations, proving the predominance of spectral cues over pitch cues in interval hearing.

By the way note that the above observation implies a possibility of continuously *modifying harmonies by timbre transformations*, similarly to pitch changes by timbre transformations in the cited experiments.

7.4   Function of Relative Hearing

Intervals, chords, and melodic lines are high-level patterns constituted by relationships between similar low-level patterns of tones. According to our model,
revealing spectral similarity is a prerequisite for recognizing audio structure.
We have shown that recognizing similarity results in a decomposition of a chord
into similar components which are associated with notes. At a higher representation level, these similar spectral patterns constitute an acoustical contour
recognized as a chord. In dynamics, the same perception mechanism provides
joining similar sounds into acoustical trajectories recognized as melodic lines.

Therefore, relative (interval) hearing can be understood as a capacity to
directly recognize the distance between tones *which are similar in spectral
structure*. It is an appearance of general capability of perception to recognize
structure by similarity, i.e. it is the correlative perception in audio domain.

Chords and melodic lines, as patterns of high-level, are invariant with respect to voice changes and pitch transpositions. Indeed, high-level patterns of
chords and melodic lines are determined by the relationships between low-level
patterns which may be not identifiable (cf. with Fig. 2.1 and Fig. 2.2 where the
contour of B is still recognizable even if A is replaced by unknown symbol II).

It follows that interval relationships which determine high-level patterns of
chords and melodic lines are more stable with respect to different factors than
low-level tone patterns characterized by pitch and timbre. Therefore, interval
relationships are fitted for carring semantic musical information better than
tones and pitch.

The above observation is often ignored by music theorists concerned with
the standard notation and its absolute pitch. However, in the past the stuff
was not provided with a pitch standard as now, and the notation has fixed
relative rather than absolute pitch. Such an interval approach to notation is
seen in the parts of *thorough bass, or figured bass,* where the harmony is written
down by magnitudes of intervals with respect to the bass line (see Fig. 7.4).

In notation a chord can be regarded as a graphical symbol drawn by notes
(cf. Fig. 2.1) which is invariant with respect to pitch transpositions. Similarly, melodic lines can be considered as trajectories invariant with respect to
transpositions; thus the graphical contour of a fugue theme in different transpositions is visually recognizable in the score.

The only imperfectness of standard notation (for visually recognizing pitch
transpositions) is the diatonic scale of the stuff, implying that some intervals
which are equal in notation are not equal in sound. For instance, the same
distance at the stuff is inherent in the following intervals: $(e_1; f_1)$ which is
equal to one semitone, and $(f_1; g_1)$ which is equal to two semitones, implying
that the graphical transposition of the first interval does not correspond to its
transposition in sound (Fig. 7.5).

Figure 7.4: Thorough bass notation based on interval relationships
(J.S.Bach's transcription of the first measures of his
aria "Empfind ich Höllenangst und Pein" (Keller 1955))

Therefore, in order to make notated chords and melodic lines invariant with respect to graphical transpositions, one has to use appropriate key signatures, corresponding to the interval of transposition (Fig. 7.6).

If the staff was graduated chromatically, a transposition would mean a graphical translation of notes with respect to the stuff. A kind of such a chromatic notation is the guitar chords fingering notation where frets correspond to semitones (Fig. 7.7). In this notation pitch transpositions precisely correspond to graphical transpositions, displaying the invariable structure of a given chord type.

Finally, note that in our model the functions of relative (interval) hearing and absolute (pitch) hearing are separated. Interval hearing is necessary at the first stage of audio data processing where it contributes to the discovery of similar audio patterns and to voice separation. Absolute hearing performs pitch identification of voices at the second stage of audio data processing, after the voices have already been separated. This strong contrast between the functions of interval and pitch hearing is an exaggeration made in order to clarify our point of view.

Thus interval hearing contributes to recognizing melodic lines and chords. In a broader sense, it tracks simultaneous audio processes. Such a capacity is extremely important for orientation in the acoustical environment, and that may be the cause of dominance of interval hearing over absolute hearing developed in evolution.

Figure 7.5: Transposition in notation but not in sound

Figure 7.6: Transposed chord and melody in standard notation

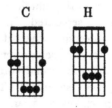

Figure 7.7: Invariance of guitar fingering with respect to chord transpositions

7.5 Counterpoint and Orchestration

In this section we add some remarks to the arguments from Section 1.4, concerning the prohibition against parallel primes, fifths, and octaves in the theory of counterpoint and discuss some other applications of our model to counterpoint and orchestration.

Recall that a *parallel voice-leading,* implying a parallel motion of partials of the voices, results in their fusion into a new voice with a new timbre. This effect is used in pipe organs where a single key activates several pipes tuned according to a certain chord.

The use of **timbral effect of parallel voice-leading** in orchestration is shown in Fig. 1.2. The theme played by horn is doubled by celesta at one and two octaves, and by flutes at the twelfth and the seventeenth. These five parallel parts are perceived as one voice instead of five. This way a new voice is synthesized with the first five harmonics enhanced by doubling. The idea of parallel voice-leading is emphasized by the key signatures in the score: The flute parts, corresponding to the third and fifth harmonics, are notated in G and E, respectively, while the fundamental tonality being C.

Thus the parallel voice-leading results rather in a timbral than a harmonic effect. In strict counterpoint this implies loss of a voice, reduction of harmony, and rise of a new timbre from the fusion of parallel voices. Therefore, the prohibition against parallel voices in counterpoint can be explained as preventing from breaking the homogeneity of polyphonic texture and providing the distinguishability of voices.

However, when the texture is sufficiently complex, the auditory effect may be timbral rather than harmonic, even if all the rules of voice-leading are observed. For example, in Fig. 7.8 the voices are not strictly parallel as in Fig. 1.2, but the complexity of the texture results in a fusion of the voices into a new voice with internal timbral fluctuations. If such a deviation from strictly parallel voice-leading was used in a more transparent texture, say as in Fig. 1.2, the harmonic effect would be immediately heard.

The perception of complex polyphony as a sonorousness was mentioned by Xenakis (1954; 1963) who has used this effect in order to justify stochastic composition instead of complex polyphony:

> Linear polyphony destroys itself by its very complexity; what one hears is in reality nothing but a mass of notes in various registers. The enormous complexity prevents the audience from following the intertwining of the lines and has as its macroscopic effect an irrational and fortuitous dispersion of sounds over the whole extent of the sonic spectrum. There is consequently a contradiction between the polyphonic linear system and the heard result, which is surface or mass. This contradiction inherent in polyphony will disappear

when the independence of sounds is total. In fact, when linear combinations and their polyphonic superpositions no longer operate, what will count will be the statistical mean of isolated states and of transformations of sonic components at a given moment. The macroscopic effect can then be controlled by the mean of the movements of elements which we select. The result is the introduction of the notion of probability, which implies, in this particular case, combinatory calculus. Here, in a few words, is the possible escape from the "linear category" in musical thought.

(The English translation is given by (Xenakis 1971, p. 8)).

According to Xenakis, a complex polyphony results in the indistinguishability of voices which is perceived as sonorousness. Since the distinguishability of voices is characterized in our model by the voice separability, the model of chord recognition can be applied to **measuring the degree of sonorousness in polyphony**. A complex polyphony whose voices are not separable (not distinguishable) should be recognized as sonorousness. If parts are separable (corresponding to the distinguishability of voices) then the polyphonic origin should be recognized as predominant. Characterizing intermediate states between good separability of voices and their total fusion by the percentage of true recognition of voices, one can quantitatively estimate the ratio of polyphonic/sonorous effects.

Our model explains the predominance of dynamic perception over static perception (Bregman 1990), or the suppression of vertical (harmony) cues by linear (melodic, polyphonic) cues. For example, each harmonic vertical in Fig. 1.2, being taken separately, can be perceived as a chord. (Even partials of a tone can be distinguished in a so called analytical mode of listening.) Nevertheless, in dynamics parallel voices fuse into one with a rich timbre. In our model this follows from the fact that recognition hypotheses based on analysis of dynamic melodic intervals between a given chord and its neighbors are numerous, in contrast to the singularity of the recognition hypothesis based on analysis of static harmonic intervals. At the stage of final decision making (see Section 5.6) this implies a majority dominance of dynamic hypotheses over the only static one, meaning the dominance of linear cues over vertical ones in recognition.

In the theory of counterpoint the principle of dominance of melodic cues over harmonic cues is expressed as **a prescription to pay more attention to logical development of voices rather than to adjusting them to consonant verticals.**

Also note that the recognition of melodic intervals in melodic lines is simpler, i.e. requires less processing and memory, if all tones of a melody are played by the same voice. In this case the correlation of related spectral patterns is maximal. Conversely, if one note is played by one instrument, the following

Figure 7.8: Sonorous effect from a complex harmony in *Bolero* by M.Ravel

note by another instrument, and so on, the correlation of spectral patterns is much less, and their linking into a melody is more difficult.

Generally speaking, the continuity of timbre is very important for the correlativity of perception, since the timbral continuity makes adjacent time-cuts similar, implying their correlation and linking into acoustical processes.

In particular, this explains the known orchestration rule of **transmitting a melody from one instrument to another through a common note**: The last note played by the first instrument is prescribed to be the first note played by the second instrument. If the second instrument enters after the last note of the first instrument, the effect is recognized as starting a new phrase. This is illustrated by Webern's orchestration of *Ricercar* from J.S.Bach's *Musical Offering* (see Fig. 7.9 and 7.10). In order to provide a fine segmentation (pointillistic) effect, the instruments do not overlap. The only exception is the transmittance of melody from horn to trombone through common note d_1 which doesn't break the phrase.

As follows from our model, linking tones into melodic lines is more reliable if their correlation is more salient. In turn tone spectra are more correlated if there are more partials in the voice spectrum. That is an explanation of the **use of bright voices (with many partials) for solo and leading voice**. As mentioned in Section 5.5, voices with a few partials are not easily recognizable in chords. Being not structurally salient, such voices interact with each other, resulting in accidental patterns and creating a uniformly dense background. This property of voices with a few partials makes them suitable for transparent "orchestral pedals", *ripieno* parts, etc.

In this connection note that the leading voice in Fig. 1.2 and Fig. 7.8 is intentionally enriched by additional parallel parts which make the theme brighter. At the same time, the upper overtones of the theme are given to flutes with their soft timbre. This is done because these parts should blend with the main voice, not being too much salient themselves. The same reason explains why organ mixture registers are also compiled from pipes with a soft timbre.

The model explains the **prohibition against large spacing** in counterpoint, i.e. against leading voices more than an octave apart, the only exception being made for bass (Fig. 7.11). Recall that in order to avoid octave autocorrelation of voices, we restrict maximal intervals considered to 12 semitones (see Section 5.3). Under such a restriction, distant tones (spaced by intervals larger than an octave) cannot be recognized, except for the lowest tone which is always recognizable by the interval of prime (see Section 5.4).

On the other hand, a distant voice which cannot be put into correlation with other voices can be recognized only as a separate trajectory but not as a voice which interacts harmonically with other parts. Such an opposition of distant voice to others is a prerequisite for using distant voices as solo parts

Figure 7.9: Beginning of *Ricercar* from J.S.Bach's *Musical Offering*

Figure 7.10: A.Webern's orchestral arrangement of J.S.Bach's *Ricercar*

Figure 7.11: Spacing between voices more than an octave prohibited in counterpoint

but not in a homogeneous polyphony.

This is a reason for the **reservation of a free vertical space for solo** in orchestration. As follows from the above paragraph, if melodic intervals of solo part do not overlap with those of the accompaniment then the separation of solo and accompaniment becomes easier. Leading a solo part above or below other voices provides a free vertical space for solo. It is well known that soprano and bass voices are perceived better than others.

Thus some rules of orchestral arrangement can be explained from the standpoint of our model. Keeping to orchestration rules, one simplifies the perception adequate to the score. On the contrary, an infringement of these rules can result in an inadequate perception of notated parts. Certainly, one can deviate from orchestration rules intentionally in order to obtain some special audio effects.

We can conclude that the rules of counterpoint are aimed at the homogeneity of polyphonic texture, whereas the rules of orchestration are aimed at the control over certain auditory effects. The polyphonic homogeneity is only one of such effects, and that is why the rules of counterpoint are not always observed in orchestration. In a sense, orchestration rules are more general. They may be reduced to the rules of counterpoint in particular cases, but in other cases they are too restrictive.

7.6   Harmony

Similarly to the previous section, the model explains some rules of harmony as simplifying music perception adequate to the score, which in our model corresponds to simplifying music recognition. Since our comments are made from a rather special standpoint, they don't pretend to cover all aspects of the phenomena discussed.

c	d	e	f	g	a	h	c_1
1	9/8	5/4	4/3	3/2	5/3	15/8	2

Figure 7.12: Frequency ratios for the scale of just intonation in *do-major*

Figure 7.13: The voices parallel in notation but not in sound

The model exhibits a certain acoustical **advantage of just intonation** over the equal temperament. Recall that in the equally tempered scale all intervals equal in notation are equal in sound. It doesn't always hold in the *scale of just intonation* which is generated by octave transpositions of tones of acoustically pure tonic, dominant, and subdominant triads. The frequency ratios of diatonic degrees to the tonic in just intonation are shown in Fig. 7.12 (Rossing 1990, p.173). Note that in just intonation the fifths of tonality C have the following frequency ratios:

- frequency ratio of fifth $(c; g)$ is equal to 2:3;

- frequency ratio of fifth $(d; a)$ is equal to $27 : 40 \neq 2 : 3$;

- frequency ratio of fifth $(e; h)$ is equal to 2:3.

This implies that the fifths in Fig. 7.13, looking parallel in notation, are not precisely parallel in sound. Since in our model every voice parallelism complicates their separation, the scale of just intonation, preventing from some parallelisms, provides the conditions for better voice separability. However, the difference between the two tunings mentioned, just intonation and equal temperament, is too small to be important. The perceptual effect is rather a different "color" of chords $(d; f; a)$ and $(e; g; h)$ in just intonation.

There are some other examples of correspondence between the rules of harmony and recognizability of voices in our model. Consider the **prohibition against voice crossing**, i.e. exchanging positions of two voices when a lower voice becomes the upper voice and vice versa (Fig. 7.14). If voices are not very much different, their crossing is usually perceived as one instrument playing the upper notes and another instrument playing the lower notes as if the voices were not crossed (McAdams & Bregman 1979). In our model the related recognition result is obtained if the intervals considered are restricted to small values, which requires less processing and memory (corresponding to easier perception).

The same reasons explain the **prohibition against voice overlapping**. Recall that voices are said to overlap if the lower voice moves above the former note of the upper voice, or the upper voice moves below the former note of the lower voice (Fig. 7.15).

Filling skips (a retrograde movement of a melody after a skip shown in Fig. 7.16) can be also justified. Indeed, filling a skip reduces the melodic interval between the chord where the skip is filled in and its second predecessor (from which the skip starts). In our model the smaller a melodic interval is, the smaller is the amount of data and processing which are required to recognize it correctly. Interpreting the simplification of recognition as the simplification of perception, we conclude that the prescription to fill skips in provides conditions for making the perception of melodies easier, i.e. making melodies more natural.

The reasons adduced here imply that it is easier to perceive a melody with small intervals (primes and seconds) than that with skips (thirds and larger intervals). In theory the corresponding voice-leading is said to be *melodic*, which emphasizes the idea of naturalness of melody, whereas a voice-leading by larger intervals is said to be *harmonic*, which indicates at some other organization of melody.

The extreme reduction of intervals between successive tones results in a **glissando** (continuous gliding from one pitch to another). As mentioned in the previous section, the principle of correlativity gives best results in recognizing continuous changes, because the correlation is greater for identical blocks of data (in the previous section we have discussed continuity in timbre; here we consider continuity in pitch). In our model a glissando is well recognizable, corresponding to easy perception. Indeed, weak sounds of Hawaiian guitar are heard distinctly even in a background of a rather loud and dense accompaniment, proving such a hypothesis.

Similarly to *glissando*, small pitch fluctuations make perceptual segregation of tones easier. Hence, we can justify the **use of ornamentation**, e.g. *vibrato*, trills, grace-notes, gliding at attacks of sounds, etc. The ornamentation, besides its coloring function, activates the perception (gives additional cues for true recognition) and attracts attention to the ornamented tones due to the recognition of their movement in a less variable background.

Such an enhancing function of ornamentation explains its wide use in harpsichord music with no other possibility to attract attention to the leading voice.

Figure 7.14: Voice crossing prohibited in counterpoint

Figure 7.15: Voice overlapping prohibited in counterpoint

Figure 7.16: Filling skips in strict counterpoint

On the contrary, the ornamentation is not used in accompaniment which must be less salient than the leading voice. It is remarkable that ornamentation is notated by special signs, other than for melody notation. This distinction indicates at a special composition function of ornamentation, rather articulatory and intonational than structural.

The model demonstrates the **fundamental role of bass voice** in harmony. As said in Section 5.4, every tone of a chord can be recognized by its correlation with the lowest tone, because the corresponding interval is not masked by any lower parallel interval. In a sense, the bass tone determines the perception of a chord, being a reference tone for all other tones.

The model reveals the **harmonic incompleteness of cluster chords** which are constituted by equal intervals (Fig. 7.17). The lower interval, shown by the tie, is repeated several times in the chord. This means that the lower interval masks the upper parallel intervals, making them less salient. As an effect, the lower interval determines the perception of the chord, since its upper tones blend with two lower tones.

In a musical context, tones of a cluster chord can be easily separated by recognizing melodic intervals between a given chord and its neighbors. In Fig. 7.17 cluster chords are shown as suspensions with resolutions. One can see that melodic intervals between any two matched chords are not parallel, enabling tone separation by melodic intervals. Therefore, in a musical context cluster chords are quite acceptable, if only they don't form parallel sequences like in Fig. 7.18. One can see that in such a case the harmonic incompleteness of cluster chords is persistent, since all melodic intervals between successive chords are parallel.

The model explains an **easy perception of major chords in the root position** (when the root of a chord is the lowest tone). In Fig. 7.19 one can see that there is no parallel melodic interval between any two fundamental triads of major key (tonic–subdominant, subdominant–dominant, and dominant–tonic). Therefore, major triads in the root position can be correctly recognized under worst conditions (in the computer experiments described in Section 5.5 the worst conditions are the following: Accuracy of spectral representation within one semitone, harmonic voices with 16 partials, and maximal intervals restricted to 12 semitones).

The same conclusion relates to fundamental triads in natural minor, but not in harmonic minor (with major dominant). For harmonic minor, the recognition of fundamental triads is complicated by parallel melodic intervals between some of the chords shown in Fig. 7.20. This observation meets the experience of music education: The beginners write musical dictations in major key better than in minor.

We see that the rules of harmony discussed are aimed at the perception adequate to the score, which in our model corresponds to the conditions pro-

Figure 7.17: Cluster chords with resolutions

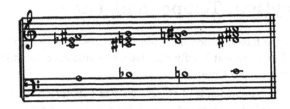

Figure 7.18: Sequence of cluster chords

Figure 7.19: Major triads in root position in harmonic and melodic junctions

Figure 7.20: Minor triads in root position in harmonic and melodic junctions

viding true recognition. An infringement of these rules causes an inadequacy of music perception, or complicates it, corresponding to recognition mistakes or the necessity of considerable amount of processing and memory for correct recognition.

Finally, note that the **interaction between musical hearing and musical memory** can be illustrated by the model. The results of chord recognition depend greatly on the accuracy of spectral representation, on the size of maximal intervals considered, as well as on the number of chords or time-cuts confronted, i.e. on the duration of the musical excerpt held in the memory. Since a better recognition reliability requires more processing and memory, we can say that the better the memory is, the better is musical hearing.

7.7   Rhythm, Tempo, and Time

In Chapter 6 we have applied our model of correlative perception to rhythm recognition. Let us repeat briefly our main conclusions concerning the nature of rhythm, tempo, and time.

According to the principle of correlativity of perception, a sequence of time events is represented in terms of generative groups of events, associated with repetitious rhythmic patterns. Repeated or slightly distorted rhythmic patterns are considered as subjective time units for tempo tracking. The "trajectory" drawn by these reference units with respect to the time axis is associated with a high-level pattern of tempo curve. Among all possible representations of data, the representation with least total complexity is chosen, while the total complexity being shared between rhythmic patterns and tempo curve.

Thus rhythm and tempo are defined as *complementary attributes of optimal representation of time data*. Rhythmic patterns are defined to be generative units of such a representation and tempo is defined by time interrelationships between them. In our model tempo and rhythm cannot be considered separately, since the two representation levels are interdependent. Moreover, minimizing the total complexity of the representation, which is shared between the two levels, we adjust rhythm to tempo and tempo to rhythm.

As shown in Section 2.2 and Chapter 6, the same sequence of time events can be interpreted either in terms of rhythm, either in terms of tempo, or both. In example from Section 2.2, the sequence of durations is interpreted first in terms of rhythm (single complex rhythmic pattern under a constant tempo) and then it is interpreted in terms of tempo (simple rhythmic pattern repeated under a tempo change), depending on the context.

Since the choice of interpretation is guided by the idea of least complexity, the optimal representation can be unstable with respect to distortions of input data, which is caused by a general instability of optima. To illustrate it, imagine an infinity of representations (e.g. different representations of time

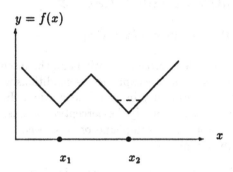

Figure 7.21: Instability of optima

events) conventionally associated with points at the axis of x in Fig. 7.21. Let the complexity of representation be a function $y = f(x)$. Suppose that the graph of this complexity function looks like in Fig. 7.21. Then representation x_2 is least complex, being optimal. Now suppose that the initial data are changed a little, resulting in a little increase in the complexity of representations which are close to x_2, as shown in Fig. 7.21 by dotted line. This is sufficient to make representation x_2 not optimal any longer (not least complex). In Fig. 7.21 this modification results in the optimality of representation x_1 which is quite distant from x_2.

Thus a slight distortion of initial data may result in a considerable change of optimal representation, implying an alternative interpretation of time events in terms of tempo and rhythm. One can observe the instability of optimal data representation with the example illustrated by Fig. 2.3 and 2.4, where the change of optimal representation is caused not even by a change of time data, but by some external factors as placing the rhythmic progression into a melodic context.

Our approach to defining the time is also influenced by the principle of correlativity of perception. Time patterns are considered as an intermediate representation level between the level of rhythmic patterns and that of tempo curve. A time pattern is defined to be a stable pre-image with respect to elaboration of generative rhythmic patterns (cf. multilevel rhythm representation in Fig. 6.1 and with the representation obtained by the end of Section 6.8).

Summing up what has been said, we conclude that three concepts, tempo, rhythm, and time correspond to three different levels of optimal time data representation. Since we optimize the representation in whole, making it least complex, the tempo, rhythm, and time are tightly interdependent. Since they determine each other, their separate recognition is not possible in the model, and we recognize them simultaneously in their interaction.

7.8   Summary of the Chapter

Let us summarize the main items of this chapter.

1. The logarithmic scale in pitch perception and the insensitivity of the ear
    to the phase of the signal are explained as conditions which are necessary
    for recognizing the causality in sound generation. In particular, these
    two properties of hearing imply the perception of musical tones as entire
    sound objects rather than compound ones, as well as the perception of
    chords as built of tones.

2. The correlativity approach to voice separation implies a new definition of
    interval. The interval is defined between two tones with similar spectra
    as the distance of translation of the corresponding spectrum along the
    $\log_2$-scaled frequency axes. Therefore, we define the interval between
    two tones with no reference to their pitch, implying the applicability of
    the definition of interval to all harmonic tones and inharmonic sounds as
    well.

3. The nature of relative (interval) hearing is understood as the correlative
    perception in the audio domain. It is supposed that interval hearing is
    an appearance of the capacity to recognize similar sound objects and
    their development in time, which is necessary for voice separation and
    tracking simultaneous acoustical processes. Therefore, interval hearing
    is not a purely music perception phenomenon, but a general perception
    mechanism.

4. Some statements of music theory are explained as conditions which sim-
    plify the perception of polyphonic music adequate to the score. In par-
    ticular, we have outlined some applications of our model to the theory
    of harmony, counterpoint, and orchestration. Deviations from certain
    rules result in special auditory effects. The consistency of statements of
    music theory with our model of correlative perception is interpreted as
    an argument in favor of its validity.

5. Three concepts related to the perception of time events, tempo, rhythm,
    and time, are defined from the standpoint of the principle of correlativity
    of perception. Rhythmic patterns are understood as reference units for
    tempo tracking, and the time is defined to be a stable preimage (with
    respect to elaboration) of generative rhythmic patterns. Tempo, rhythm,
    and time are defined interdependently, as certain elements of optimal
    representation of time data.

Chapter 8

General Discussion

The applicability of the proposed model to two different problems, voice separation and tempo tracking, supports the validity of the hypothesis about the existence of correlative perception in humans. Moreover, the existence of correlative perception is indirectly substantiated by the consistency of the proposed model with music theory. So many coincidences are very unlikely by chance alone.

It is reasonable too that the perception reduces the redundancy of the input information in order to achieve a compact representation. Most likely, the brain tends to save the memory store and facilitates access to accumulated knowledge by representing information in an aggregate form.

As mentioned in Chapter 1 and Section 7.2, musical signal is a carrier of some semantic information. The task of music recognition is therefore extracting this information which can be realized by data aggregation. This information falls into several categories. A listener perceives structurally organized sound, fine execution nuances, emotions of the performer, and acoustic characteristics of instruments and environment. For example, performer's emotions (joy, anger, etc.) are transmitted by so-called essentic curves in loudness and tempo changes (Clynes 1977; 1983). A remarkable peculiarity of timbre to represent physical phenomena (force, tension, etc.) is described by Cadoz (1991). We can say that most semantic musical information relates to different forms of causality in sound.

Our study is devoted to the recognition of special cases of causality, structural causality. We adduce arguments in favor of the fact that an appropriate representation contributes to the recognition of audio structure by identity and by similarity, both in pitch and time domains.

In particular, we deal with the decomposition of chords into tones. From a physical standpoint, all sounds are decomposable into pure tones which correspond to resonances of vibrating bodies. Consequently, from a physical point of view a chord cannot be regarded as more complex than a monolithic complex tone. Indeed, both can be produced by a single vibrating body (loudspeaker, piano board), whereas tones are perceived as entireties, and chords are perceived as compounds. On the other hand, sounds from several physical sources can fuse into one, as in case of voice synthesis in pipe-organ and symphony orchestra (see Sections 1.4 and 7.5).

Therefore, the difference between monolithic sounds and sound complexes relates to psychology, information theory, and computer science rather than to physical acoustics. In fact, our goal is to find groups of audio data whose complexity is intermediate between that of sinusoidal tones and spectra of complex sounds, i.e. we deal with representations of data.

In our study a sound is said to be compound if its spectrum can be structured into independent groups of partials. The grouping is performed with respect to the similarity of groups and with respect to the criterion of least complex total representation. In dynamics this representation links similar spectra into acoustical trajectories, contributing to segregation of acoustical processes (polyphonic voices). In statics we find constituents of sound complexes (notes). We prove that optimal representation of a tone reveals its monolithic nature but optimal representation of a chord does reveal its compound structure.

We see that optimal data representation reconstructs physical causality in sound generation, revealing several excitation sources which result in a sensation of a chord. It is remarkable that the two different matters, physical causality and optimality in data representation, correspond to each other.

We can adduce some general arguments in favor of this correspondence. Since most physical processes evolve continuously, their successive states are not very much different from each other. Usually, these successive states have certain trends which are determined by some causes. On the other hand, these causes are usually not numerous, implying the effects to be not numerous too, so that the trends cannot mix chaotically. Therefore, the classification of data with respect to different trends results in the data representation where the segregated trends correspond to certain causes. Since the reaction of a physical system to an excitation is in a sense "optimal", the corresponding description should be "optimal" too. Therefore, finding the optimal description of a process is a way to recognize causal relations.

The next question is how the trends mentioned can be recognized. If we assume the continuity of physical processes, the trends can be revealed as corresponding to minor changes in the successive states of the phenomenon. To detect these minor changes, correlation analysis of slightly distorted data together with the method of variable resolution are quite suitable.

As mentioned in Section 2.4, both correlation analysis and method of variable resolution are realizable on neuron nets with parallel computing. This is consistent with the hypothesis that similar functions can be performed by the brain (Rossing 1990, p. 164), indirectly justifying our approach to perception modeling.

Thus we formulate the following hypothesis.

Data representation in terms of generative elements and their transformations reveals certain aspects of physical causality in sound generation. Such a representation is inherent in human perception, enabling source separation and tracking simultaneous audio processes.

The capacity to separate sounds and track simultaneous audio processes is extremely important for orientation in the environment and for semantic organization of information. From our point of view, the importance of these tasks explains the predominance of related audio mechanisms in audio perception (predominance of relative hearing over absolute hearing).

We suppose that our approach to perception modeling is quite general. In fact, we have already implemented models of chord recognition and tempo tracking, and we expect that the same model can be adapted to some other purposes.

In particular, the model may be applied to speech recognition. To recognize phonemes it is often proposed to recognize the contribution of different parts of the voice tract to the resulting speech signal. Since the parts of the voice tract has their own acoustical characteristics, we can pose the problem as a separation of "polyphonic lines" in the speech signal, where each line is associated with a certain part of the voice tract. Therefore, to recognize a phoneme, one needs to recognize the "chord" produced by the activated parts of the voice tract. Our model for polyphony tracking seems to be adaptable for that purpose.

Our approach can be extended from unidimensional data arrays (as audio spectra or sequences of time events) to two-dimensional data arrays with possible applications to image processing. It can be used for object separation in dynamics (by analogy with voice separation) and contour recognition (by analogy with chord recognition) in visual scene analysis.

For example, to segregate a moving object in dynamics, one can perform correlation analysis of successive instant images. In order to find both the object and its trajectory, it is necessary to find the deformations of the images which provide their high correlation. This can be done by the method of variable resolution (see Sections 2.3 and 2.4).

Further applications concern modeling of abstract thinking; see Giunchiglia & Walsh (1992) for a review. Note that our approach is based on revealing stable relationships between data blocks. Regarding stable relationships as new data, one can reveal stable relationships between stable relationships, etc. This way we come to abstract concepts which correspond to stable invariants of data representation.

Note that meaning can be understood as identifiable associations between memory patterns. Then an aggregate form of data which reveals stable relationships is the first step towards semantic organization of information.

A multi-level generalization of the model is imaginable where every next level of patterns is formed by stable relationships between patterns of lower level. To recognize meaning, the proposed model should be provided with associations between perceptual patterns (configurations of different levels) and memory patterns acquired in a previous experience. By interfacing such a model to a data base and extending to this base the methods for discovering correlations, a cognitive model can be obtained. In this model, abstract concepts are formed from associations between the patterns of high and higher levels. In other words, semantic analysis is understood as information analysis based on data aggregation and discovering the correlations of the aggregates.

In dealing with interactions of the model with a data base, the most promising solution is to use them jointly. Memory patterns (configurations of the data base) may compete with perceptual patterns (configurations of the artificial perception model), as when current patterns are compared immediately with those from a previous experience.

In this case the optimal representation of current data (in artificial perception model) may not be optimal in the joint model, i.e. the interpretation of a current message can be simpler in terms of memory patterns. Obviously, the set of memory patterns influences the system performance; it is to be expected that the different responses of different individual humans are conditioned by different experience, learning histories, or expectancies.

This means that the model of correlative perception complements the methodology of artificial intelligence with a stage of *"artificial perception"* which operates at the data input. If used in a proper artificial intelligence domain, the present model can achieve self-organization of knowledge. In pattern recognition, the model can be used to separate patterns, thereby making their identification easier. Therefore, artificial perception can interact with artificial intelligence in various modes. This corresponds to the interaction of human perception and human intelligence which complement each other and influence reciprocally.

In conclusion we point out that we may exaggerate the role of correlative perception, which operates side-by-side with many other perception mechanisms. The proposed mathematical model of correlative perception which is based on correlation analysis should not be regarded as universal or complete. To us, it is the idea of representing data optimally in terms of generative elements and their transformations which seems rather general.

Finally, we point out that neither the experiments, nor the theoretical rationale which have been put forth here should be seen as final ends. Rather this work is less a summary of theoretical applications and the obtained results than it is a posing of new problems.

Chapter 9

Conclusions

1. We develop an artificial perception approach to pattern recognition. The problem is to recognize structure by identity and structure by similarity. After the structure has been recognized and its elements have been segregated, their recognition is much easier. The model is based on a so called principle of correlativity of perception which is a joint use of grouping and simplicity principles known in psychology. The idea of the correlativity principle is finding a multilevel grouping of stimuli with least total complexity, where the complexity is understood in the sense of Kolmogorov, as the amount of memory store needed for the data representation.

2. The mathematical model considered is based on self-organizing data by their representation in terms of generative elements and their transformations, while subordinating the total representation to the criterion of least complex representation. This is realized by correlation analysis of data under their transformations. In order to perform a directional search for the data transformations which provide high correlation of similar messages, the method of variable resolution is proposed.

3. The model is applied to the problem of chord recognition. A power spectrum of a chord is represented as generated by a spectrum of a tone which is translated along the $\log_2$-scaled frequency axis. We prove the optimality of such a representation of spectral data, corresponding to the causality in the data generation. For this purpose a special mathematical machinery is developed, the deconvolution of a spectrum into irreducible spectra, similarly to the factorization of integers into primes.

4. A simplified model better suitable for practical purposes is substantiated. Instead of power spectra we consider Boolean spectra which are more stable with respect to pitch transpositions of tones. We formulate a necessary condition for generative tone patterns which justifies the use of correlation analysis for discovering generative tone patterns.

5. The model is tested on a series of computer experiments. Since the recognition of intervals is the main procedure in our model of chord recognition, the performance of the algorithm is firstly investigated for the recognition of chords by intervals. In our experiments we use synthesized chords taken from J.S.Bach's four-part polyphony. In particular, we examine different types of recognition mistakes, efficiency and stability of the recognition procedure, and decision making approach based on confronting several hypotheses about the chord obtained from analysis of different types of intervals. Under the worst conditions the reliability of recognition is about 98%.

6. Secondly, we investigate a structural approach to chord recognition where the interval structure of a chord is recognized simultaneously by multi-autocorrelation analysis of chord spectrum and finding optimal spectral representations. This approach is tested on recognizing four-part and five-part synthesized chords. It is shown that the recognition capacity of the model is close to the limits of human perception and resembles its properties: The chord recognizability is better for voices with fewer partials and for more accurate spectral representation, corresponding to better sharpness of music hearing. The harmonies are recognized best, next goes the chord structure, and then chord tones and their pitch.

7. The model is applied to the problem of rhythm recognition. A series of time events is represented in terms of transformations of generative rhythmic patterns, while their temporal distortions being associated with the tempo curve. In order to substantiate a directional search for generative rhythmic patterns, a kind of rhythmic grammar is developed. We formulate segmentation rules based on timing accentuation with which rhythmic syllables are determined. Then we define elaboration, sum, and junction of rhythmic syllables which are used in the structural representation of a sequence of time events. The performance of the algorithm is traced with an example of recognizing the structure of a given rhythm and determining its time.

8. Some applications of the model to psychoacoustics are outlined. In particular, we show that logarithmic scale in pitch perception and insensitivity of the ear to the phase of the signal are necessary for the perception of musical tones as indecomposable sound objects. Besides, we explain the nature of interval hearing as the appearance of correlative perception in audio domain. We show that interval hearing is based on recognition of audio structure by similarity and estimation of distance between similar components. In other words, interval hearing results from the capacity to separate and track simultaneous acoustical processes.

9. The model implies a refinement of definitions of interval, chord, and melodic line. An interval is defined to be the translation distance between similar spectral patterns. A chord is understood as a contour drawn by a spectral pattern of tone in the frequency domain. A melodic line is understood as a trajectory drawn by a spectral pattern of tone in time. Since spectral similarity is the only condition for linking tones in intervals, chords, and melodic lines, their traditional definitions in terms of pitch are generalized to voices with no pitch salience.

10. The model implies a refinement of definitions of rhythm, tempo, and time. The enumerated concepts are associated with different levels of the representation of time events in the model of correlative perception. They are defined in interaction, corresponding to the interaction of the levels of the model. The ambiguity in their determination (rhythm is defined with respect to tempo, and tempo is defined with respect to rhythm) is overcome by optimizing the total complexity of the data representation.

11. The model explains some statements of music theory as conditions aimed at simplifying music perception adequate to the score. From the standpoint of the model, some empirical rules of counterpoint, harmony, and orchestration are rationalized. In particular, we analyze the reasons why certain rules of counterpoint are incompatible with that of orchestration.

12. Finally, we discuss further applications of the model of correlative perception to computer vision, speech recognition and modeling of abstract thinking.

References

Aldwell E. & Schachter C. (1978) *Harmony and Voice Leading. Vol. 1.* New York: Harcourt Brace Jovanovich.

Allen P.E. & Dannenberg R.B. (1990) Tracking Musical Beats in Real Time. *Proceedings of the International Computer Music Conference'1990.* San Francisco: Computer Music Association, 140–143.

Aloimonos J. & Shulman D. (1989) *Integration of Visual Modules.* Boston: Academic Press.

Ashton A. (1971) *Electronics, Music, and Computers.* Salt Lake City: University of Utah, Computer Science Department, UTEC-CSc-71-117.

Askenfelt A. (1976) Automatic Notation of Played Music. Stockholm: Royal Institute of Technology, Report STL–QPSR 1/1976, 1–11.

Attneave F. (1954) Some Informational Aspects of Visual Perception. *Psychological Review,* 61, 183–193.

Attneave F. (1982) Prägnanz and Soap-Bubble Systems: A Theoretical Explanation. In: Beck J. (Ed.) *Organization and Representation in Perception.* Hillside, New Jersey: Erlbaum.

Bamberger J. (1980) Cognitive Structuring in the Apprehension of Simple Rhythms. *Archives de Psychologie,* 48, 171–199.

Barlow H.B. (1972) Single Units and Sensation: A Neuron Doctrine for Perceptual Psychology? *Perception,* 1, 371–394.

Barlow H.B., Narasimmhan R., & Rosenfeld A. (1972) Visual Pattern Analysis in Machines and Animals. *Science,* 177, 567–575.

Barrow H.G. & Tenenbaum J.M. (1993) Retrospective on "Interpreting Line Drawings as Three-Dimensional Surfaces." *Artificial Intelligence,* 59, 71–80.

Benade A.H. (1976) *Fundamentals of Musical Acoustics.* New York: Oxford University Press.

Bergevin R. & Levine M. (1993) Generic Object Recognition: Building and Matching Coarse Descriptions from Line Drawings. *IEEE Transactions on Pattern Analysis and Machine Intelligence*, 15(1), 19–36.

Berlioz H. (1855) *Grand traité d'instrumentation et d'orchestration moderne.* Paris: Schonenberger.

Birkhoff G. & Mac Lane S. (1965) *A Survey of Modern Algebra.* New York: MacMillan.

Blake A. & Marinos C. (1990) Shape from Texture: Estimation, Isotropy and Moments. *Artificial Intelligence*, 45, 323–380.

Bolles R.S. & Cain R.A. (1982) Recognizing and Locating Partially Visible Objects: The Local-Feature-Focus Method. *International Journal of Robotics Research*, 1(3), 57–82.

Boroda M.G. (1985) On Some Rules of Rhythmic Recurrence in Folk and Professional Music. *Kompleksnoye Izutcheniye Muzykalnogo Tvortchestva: Kontzepziya, Problemy, Perspektivy.* Tbilisi: Nauka, 135–167. (Russian).

Boroda M.G. (1988) Towards the Basic Semantic Units of a Musical Text. *Musikometrika*, 1. Bochum: Brockmeyer, 11–68.

Boroda M.G. (1991) The Concept of "Metrical Force" in Music with Bar Structure. *Musikometrika*, 3. Bochum: Brockmeyer, 59–94.

Bouman Ch. (1991) Multiple Resolution Segmentation of Textural Images. *IEEE Transactions on Pattern Analysis and Machine Intelligence*, 13(2), 99–113.

Bregman A.S. (1990) *Auditory Scene Analysis: The Perceptual Organization of Sound.* Cambridge, Massachusetts: M.I.T. Press.

Brooks R. (1981) Symbolic Reasoning Among 3-Dimensional Models and 2-Dimensional Images. *Artificial Intelligence*, 17, 285–349.

Cadoz C. (1991) Timbre et causalite. In: Barrière J.-B. (ed.) *Le timbre. Métaphore pour la composition.* Paris: IRCAM, Christian Bourgois Editeur, 17–46.

Calude C. (1988) *Theories of Computational Complexity.* Amsterdam: North-Holland.

Chafe C. & Jaffe D. (1986) Source Separation and Note Identification in Polyphonic Music. *Proceedings of the IEEE-IECEJ'ASJ International Conference on Acoustics, Speech, and Signal Processing, Tokyo, 1986*, 1289–1292.

Chafe C., Jaffe D., Kashima K., Mont-Reynaud B., & Smith J. (1985) Techniques for Note Identification in Polyphonic Music. *Proceedings of the International Computer Music Conference'1985.* San Francisco: Computer Music Association, 399–405.

Chafe C., Mont-Reynaud B., & Rush L. (1982) Toward an Intelligent Editor of Digital Audio: Recognition of Musical Constructs. *Computer Music Journal,* 6(1), 30–41.

Chateauneuf A. (1993) *Letter to A.Tanguiane.* Paris, 7th April 1993.

Clarke E. (1987) Categorical Rhythm Perception, an Ecological Perspective. In: Gabrielsson A. (Ed.) *Action and Perception in Rhythm and Music.* Stockholm: Publication of Royal Swedish Academy of Music No. 55.

Clarke E. & Krumhansl C.L. (1990) Perceiving Musical Time. *Music Perception,* 7, 213–252.

Clynes M. (1977) *Sentics: The Touch of Emotions.* New York: Doubleday Anchor.

Clynes M. (1983) Expressive Microstructure in Music Linked to Living Qualities. In: Sundberg J. (Ed.) *Studies in Music Performance.* Stockholm: Publication of Royal Swedish Academy of Music No. 39, 76–181.

Dannenberg R.B. & Bookstein K. (1991) Practical Aspects of a MIDI Conducting Program. *Proceedings of the International Computer Music Conference'1991.* Montreal: Faculty of Music, McGill University, 537–540.

Dannenberg R.B. & Mont-Reynaud B. (1987) Following an Improvisation in Real Time. *Proceedings of the International Computer Music Conference'1987.* San Francisco: Computer Music Association, 241–248.

Darwin C.J. (1984) Perceiving Vowels in the Presence of Another Sound: Constants on Formant Perception. *Journal of the Acoustical Society of America,* 76, 1636–1647.

Desain P. (1992) A (De)Composable Theory of Rhythm. *Music Perception,* 9(4).

Desain P. & Honing H. (1989) Quantization of Musical Time: A Connectionist Approach. *Computer Music Journal,* 13(3), 56–66.

Desain P. & Honing H. (1991) Tempo Curves Considered Harmful. *Proceedings of the International Computer Music Conference'1991.* Montreal: Faculty of Music, McGill University, 143–149.

Desain P., Honing H., & de Rijk K. (1989) A Connectionist Quantizer. *Proceedings of the International Computer Music Conference'1989.* San Francisco: Computer Music Association, 80–85.

Dumesnil R. (1979) *Le Rythme musical.* Paris–Geneve: Slatkine Reprints, serie "Ressources".

Ellis D. & Vercoe B. (1991) A Wavelet-Based Sinusoid Model of Sound for Auditory Signal Separation. *Proceedings of the International Computer Music Conference'1991.* Montreal: Faculty of Music, McGill University, 86–89.

Fischer M.A. & Bolles R.C. (1983) Perceptual Organization and the Curve Partitioning Problem. *Proceedings of the 8th Joint Conference on Artificial Intelligence'83, Karlsruhe,* 1014–1018.

Foster S., Schloss W.A., & Rockmore A.J. (1982) Toward an Intelligent Editor of Digital Audio: Signal Processing Methods. *Computer Music Journal,* 6(1), 42–51.

Fraisse P. (1983) Rhythm and Tempo. In: Deutsch D. (Ed.) *The Psychology of Music.* London: Academic Press.

Frances R. (1972) *La perception de la musique.* 2nd ed. Paris: Librairie philosophique J.Vrin.

Freeman W.J. (1979) EEG Analysis Gives Models of Neuronal Template Matching Mechanism for Sensory Search with Olfactory Bulb. *Biological Cybernetics,* 35, 221–234.

Friedland N.S. & Rosenfeld A. (1992) Compact Object Recognition Using Energy-Function-Based Optimization. *IEEE Transactions on Pattern Analysis and Machine Intelligence,* 14(7), 770–777.

Gelfand S.A. (1981) *Hearing. An Introduction to Psychological and Physiological Acoustics.* New-York–Basel: Marcel Dekker.

Gibson J.J. (1950) *The Perception of the Visual World.* Boston: Houghton Mifflin.

Gibson J.J. (1966) *The Senses Considered as Perceptual Systems.* Boston: Houghton Mifflin.

Gibson J.J. (1979) *The Ecological Approach to Visual Perception.* Boston: Houghton Mifflin.

Giunchiglia F. & Walsh T. (1992) A Theory of Abstraction. *Artificial Intelligence,* 57, 323–389.

Gong S. & Buxton H. (1992) On the Visual Expectations of Moving Objects. *Proceedings of the 10th European Conference on Artificial Intelligence'92, Vienna.* Wiley: Chichester, 781–784.

Gut S. (1976) Intervalle. In: Honnegger M. (Ed.) *Dictionnaire de la Musique. Science de la Musique. Formes, Technique, Instruments. Vol. 1.* Paris: Bordas.

Handel S. (1989) *Listening: An Introduction to the Perception of Auditory Events.* Hillside, New Jersey: Erlbaum.

Helmholtz H. (1877) *On the Sensation of Tone as a Psychological Basis for the Study of Music.* 4th ed. English transl.: New York: Dover, 1954.

Hewlett H.B. & Selfridge-Field E. (Eds.) (1990) *Computing in Musicology. A Directory of Research.* Menlo Park, California: Center for Computer-Assisted Research in the Humanities.

Hewlett H.B. & Selfridge-Field E. (Eds.) (1991) *Computing in Musicology. A Directory of Research.* Menlo Park, California: Center for Computer-Assisted Research in the Humanities.

Hochberg J. & McAlister E. (1953) A Quantitative Approach to Figural "Goodness". *Journal of Experimental Psychology,* 46, 361–364.

Hoffmann A.G. (1992) Phenomenology, Representations and Complexity. *Proceedings of the 10th European Conference on Artificial Intelligence'92, Vienna.* Wiley: Chichester, 610–614.

Horn B.K.P. (1975) Obtaining Shape from Shading Information. In: Winston P.H. (Ed.) *The Psychology of Computer Vision.* New York: McGraw-Hill.

Horn B.K.P. (1986) *Robot Vision.* New York: McGraw-Hill.

Horn B.K.P. & Schunck B.G. (1981) Determining Optical Flow. *Artificial Intelligence,* 17, 185–203.

Horn B.K.P. & Schunck B.G. (1993) "Determining Optical Flow": A Retrospective. *Artificial Intelligence,* 59, 81–87.

Horn B.K.P. & Weldon Jr. (1988) Direct Methods for Recovering Motion. *International Journal of Computer Vision,* 2(1), 51–76.

Hummel R. (1987) The Scale-Space Formulation of Pyramid Data Structures. In: Uhr L. (Ed.) *Parallel Computer Vision.* Boston: Academic Press.

Ikeuchi K. (1993) Comment on "Numerical Shape from Shading and Occluding Boundaries." *Artificial Intelligence*, 59, 89–94.

Ikeuchi K. & Horn B.K.P. (1981) Numerical Shape from Shading and Occluding Boundaries. *Artificial Intelligence*, 17, 141–184.

Kanade T. (1993) From a Real Chair to a Negative Chair. *Artificial Intelligence*, 59, 95–101.

Kanatani K. (1984) Detection of Surface Orientation and Motion from Texture by a Stereological Technique. *Artificial Intelligence*, 23, 213–237.

Kanatani K. & Chou T. (1989) Shape from Texture: General Principle. *Artificial Intelligence*, 38, 1–48.

Kass M., Witkin A., Terzopoulos D., & Barr A. (1988) Snakes: Active Contour Models. *International Journal of Computer Vision*, 1, 321–331.

Katayose H. & Inokuchi S. (1989a) The Kansei Music System. *Computer Music Journal*, 13(4), 72–77.

Katayose H. & Inokuchi S. (1989b) An Intelligent Transcription System. *Proceedings of the First International Conference on Music Perception and Cognition, Kyoto, Japan, 17–19 October, 1989*, 95–98.

Katayose H. & Inokuchi S. (1990) The Kansei Music System'90. *Proceedings of the International Computer Music Conference'1990*. Glasgow, 308–310.

Katayose H., Kato H., Imai M., & Inokuchi S. (1989) An Approach to an Artificial Music Expert. *Proceedings of the International Computer Music Conference'1989*. San Francisco: Computer Music Association, 139–146.

Katuar G. (1926) *Muzykalnaya Forma. 1. Ritm.* Moscow. (Russian).

Keller H. (1955) *Schule des Generalbass-spiels*. Kassel:Bärenreiter-Ausgabe 490.

Kendall G.S. & Martens W.L. (1984) Simulating the Cues of Spatial Hearing in Natural Environment. *Proceedings of the International Computer Music Conference'1984*. San Francisco: Computer Music Association, 111–125.

Kendall G.S., Martens W.L., Freed D.J., Ludwig M.D., & Karstens R.W. (1986) Simulating the Cues of Spatial Hearing in Natural Environment. *Proceedings of the International Computer Music Conference'1986*. San Francisco: Computer Music Association, 285–292.

Knowlton P.H. **(1971)** *Interactive Communication and Display of Keyboard Music.* Ph.D. Dissertation. Salt Lake City: University of Utah.

Knowlton P.H. **(1972)** Capture and Display of Keyboard Music. *Datamation*, May.

Kolmogorov A.N. **(1965)** Three Approaches to Defining the Notion "Quantity of Information". *Problemy Peredatchi Informatsii*, 1(1), 3–11. Reprinted in: Kolmogorov A.N. *Theory of Information and Theory of Algorithms.* Moscow: Nauka, 1987, 213–223. (Russian).

Leeuwenberg E.L.J. **(1971)** A Perceptual Coding Language for Visual and Auditory Patterns. *American Journal of Psychology*, 84, 307–350.

Leeuwenberg E.L.J. **(1978)** Quantification of Certain Visual Properties: Salience, Transparency, Similarity. In: Leeuwenberg E.L.J. & Buffart H.F.J.M. (Eds.) *Formal Theories in Visual Perception.* Chichester UK: Wiley.

Lerdahl F. & Jackendoff R. **(1983)** *A Generative Theory of Tonal Music.* Cambridge, Massachusetts: M.I.T. Press.

Leyton M. **(1986)** Generative Systems of Analyzers. In: Rosenfeld A. (Ed.) *Human and Machine Vision II.* Orlando: Academic Press, 149–189.

Lifshitz L.M. & Pizer S.M. **(1990)** A Multi-Resolution Hierarchical Approach to Image Segmentation Based on Intensity Extrema. *IEEE Transactions on Pattern Analysis and Machine Intelligence*, 12(3), 234–254.

Longuet-Higgins H.C. **(1976)** The Perception of Melodies. *Nature*, 263, 646–653.

Longuet-Higgins H.C. **(1987)** *Mental Processes.* Cambridge, Massachusetts: M.I.T. Press.

Longuet-Higgins H.C. & Lee C.S. **(1982)** The Perception of Music Rhythms. *Perception*, 11, 115–128.

Longuet-Higgins H.C. & Lee C.S. **(1984)** The Rhythmic Interpretation of Monophonic Music. *Music Perception*, 1, 424–441.

Longuet-Higgins H.C. & Prazdny K. **(1984)** The Interpretation of a Moving Retinal Image. *Proceedings of the Royal Society*, B208, 385–397.

Marr D. **(1982)** *Vision.* San Francisco: W.H.Freeman.

Marr D. & Nishihara H.K. **(1978)** Representation and Recognition of the Spatial Organization of Three-Dimensional Shapes. *Proceedings of the Royal Society*, B200, 269–291.

Marr D. & Poggio T. (1976) Cooperative Computation of Stereo Disparity. *Science*, 194, 283–287.

Martens W.L. (1987) Principal Components Analysis and Resynthesis of Spectral Cues to Perceived Direction *Proceedings of the International Computer Music Conference'1987*. San Francisco: Computer Music Association, 274–281.

McAdams S. (1989) Segregation of Concurrent Sounds. I: Effects of Frequency Modulation Coherence. *Journal of the Acoustical Society of America*, 86, 2148–2159.

McAdams S. (1991a) Segregation of Concurrent Sounds. II: Effects of Spectral Envelope Tracing, Frequency Modualation Coherence, and Frequency Modulation Width. *Journal of the Acoustical Society of America*, 89, 341–351.

McAdams S. (1991b) Perceptual Organization of Audio Environment. In: Nosulenko V. (Ed.) *Problems of Ecological Psychoacoustics*. Moscow: Institute of Psychology of the Academy of Sciences, 28–50. (Russian).

McAdams S. (1993) Recognition of Sound Sources and Events. In: Bigand E. & McAdams S. (Eds.) *Thinking in Sound: Cognitive Perspectives on Human Audition*. Oxford: Oxford University Press, 146–198.

McAdams S. & Bregman A. (1979) Hearing Musical Streams. *Computer Music Journal*, 3(4), 26–43, 60, 63. Reprinted in: Roads C. & Strawn J. (Eds.) *Foundations of Computer Music*. Cambridge, Massachusetts: M.I.T. Press, 1985, 658–698.

Mellinger D. & Mont-Reynaud B. (1991) *Proceedings of the International Computer Music Conference'1991,* Montreal: Faculty of Music, McGill University, 90–93.

Messiaen O. (1944) *Technique de mon langage musical. Vol. 1.* Paris: Leduc.

Meygret A. & Thonnat M. (1990) Segmentation of optical Flow and 3D Data for the Interpretation of Mobile Objects. *Proceedings of the 3rd International Conference on Computer Vision, Osaka.* 238–245.

Michon J.A. (1964) Studies on Subjective Duration. *Acta Psychologica*, 222, 441–450.

Milner P.N. (1974) A Model for Visual Shape Recognition. *Psychological Review*, 81(6), 521–535.

Minsky M. & Papert S. (1988) *Perceptrons.* 2nd ed. Cambridge, Massachusetts: M.I.T. Press.

Minsky M. (1975) A Framework for Representing Knowledge. In: Winston P.H. (Ed.) *The Psychology of Computer Vision.* New York: McGraw-Hill.

Mohan R. & Navatia R. (1992) Perception of 3-D Surfaces from 2-D Contours. *IEEE Transactions on Pattern Analysis and Machine Intelligence,* 14(6), 616–635.

Mont-Reynaud B. (1985) Problem-Solving Strategies in a Music Transcription System. *Proceedings of the 9th International Joint Conference on Artificial Intelligence.* Los Angeles, 916–918.

Mont-Reynaud B. & Goldstein M. (1985) On Finding Rhythmic Patterns in Musical Lines. *Proceedings of the International Computer Music Conference'1985.* San Francisco: Computer Music Association, 391–397.

Mont-Reynaud B. & Gresset E. (1990) PRISM: Pattern Recognition in Sound and Music. *Proceedings of the International Computer Music Conference'1990.* Glasgow, 153–155.

Mont-Reynaud B. & Mellinger D. (1989) Source Separation by Frequency Co-Modulation. *Proceedings of the First International Conference on Music Perception and Cognition, Kyoto, Japan, 17–19 October, 1989,* 99–102.

Moore B.C.J. (1982) *Introduction to the Psychology of Hearing.* 3rd ed. London: Academic Press.

Moore F.R. (1978) An Introduction to the Mathematics of Digital Signal Processing. *Computer Music Journal,* 2(1), 38–47; 2(2), 38–60. Reprinted in: Strawn J. (Ed.) *Digital Audio Signal Processing: An Anthology.* Madison: A-R Edition, 1985, 1–67.

Moorer J.A. (1975) *On the Segmentation and Analysis of Continuous Musical Sound by Digital Computer.* Ph.D. Thesis. Stanford: Stanford University, Dep. of Music Report STAN-M-3.

Moorer J.A. (1977) On the Transcription of Musical Sound by Computer. *Computer Music Journal,* 1(4), 32–38.

Morita S., Kawashima T., & Aoki Y. (1992) Hierarchical Shape Recognition Based on 3-D Multiresolution Analysis. In: Sandini G. (Ed.) *Computer Vision – ECCV'92.* Lecture Notes in Computer Science 588. Berlin: Springer, 843–851.

Moses Y. & Ullman S. (1992) Limitations of Non-Model-Based Recognition Schemes. In: Sandini G. (Ed.) *Computer Vision – ECCV'92.* Lecture Notes in Computer Science 588. Berlin: Springer, 820–828.

Navab N. & Zhang Z. (1992) From Multiple Objects Motion Analysis to Behavior-Based Object Recognition. *Proceedings of the 10th European Conference on Artificial Intelligence'92, Vienna.* Wiley: Chichester, 790–794.

Neisser U. (1966) *Cognitive Psychology.* New York: Appelton-Century-Crofts.

Niihara T. & Inokuchi S. (1986) Transcription of Sung Song. *Proceedings of the International Conference on Acoustics, Speech, and Signal Processing'86, Tokyo, April 7–11, 1986,* 1277–1280.

Oppenheim A.V. & Schafer R.W. (1975) *Digital Signal Processing.* Englewood Cliffs, New Jersey: Prentice-Hall.

Palmer S.E. (1975) Visual Perception and World Knowledge: Notes on a Model of Sensory-Cognitive Interaction. In: Norman D.A., et al. (Eds.) *Exploration in Cognition.* Hillsdale, New Jersey: Erlbaum.

Palmer S.E. (1982) Symmetry, Transformation, and the Structure of Perceptual Systems. In: Beck J. (Ed.) *Organization and Representation in Perception.* Hillsdale, New Jersey: Erlbaum.

Palmer S.E. (1983) The Psychology of Perceptual Organization: A Transformational Approach. In: Beck J., Hope B., & Rosenfeld A. (Eds.) *Human and Machine Vision.* New York: Academic Press, 269–339.

Pentland A. & Horowitz B. (1991) Recovery of Nonrigid Motion and Structure. *IEEE Transactions on Pattern Analysis and Machine Intelligence,* 13(7), 730–742.

Pistone D. (1977) Tempo. In: Honnegger M. (Ed.) *Dictionnaire de la Musique. Science de la Musique. Formes, Technique, Instruments.* Paris: Bordas.

Piszczalski M. & Galler B.A. (1977) Automatic Music Transcription. *Computer Music Journal,* 1(4), 24–31.

Piszczalski M., Galler B.A., Bossemeyer R., Hatamian M., & Looft F. (1981) Performed Music: Analysis, Synthesis, and Display by Computer. *Journal of the Audio Engeneering Society,* 29(1/2), 38–46.

Pitts W. & McCulloch W.S. (1947) How we Know Universals: The Perception of Auditory and Visual Forms. *Bulletin of Mathematical Biophysics*, 9, 127–147.

Pollard H. (1950) *The Theory of Algebraic Numbers.* Baltimore: Mathematical Association of America/Wiley.

Porte D. (1977) Rythme. In: Honnegger M. (Ed.) *Dictionnaire de la Musique. Science de la Musique. Formes, Technique, Instruments.* Paris: Bordas.

Posner M.I. (1978) *Chronometric Explorations of Mind.* Hillside, New Jersey: Erlbaum.

Posner M.I. & Henik A. (1983) Isolating Representational Systems. In: Beck J., Hope B., & Rosenfeld A. (Eds.) *Human and Machine Vision.* New York: Academic Press, 481–543.

Povel D.J. & Essens P. (1985) Perception of Temporal Patterns. *Music Perception*, 2(4), 411–440.

Preparata F.P. & Shamos M.I. (1985) *Computational Geometry.* New York: Springer.

Pridmore T.P., Mayhew J.E.W., & Frisby J.P. (1990) Exploiting Image-Plane Data in the Interpretation of Edge-Based Binocular Disparity. *Computer Vision, Graphics, and Image Processing*, 52, 1–25.

Rabiner L.R. & Gold B. (1975) *Theory and Applications of Digital Signal Processing.* Englewood Cliffs, New Jersey: Prentice-Hall.

Rangarajan K. & Shah M. (1991) Establishing Motion Correspondence. *Computer Vision, Graphics, and Image Processing: Image Understanding*, 54(1), 56–73.

Raynaut F. & Samuel A. Oriented Shift of Representational Bias for Elementary Patterns. *Proceedings of the 10th European Conference on Artificial Intelligence'92, Vienna.* Wiley: Chichester, 487–489.

Rédei L. (1967) *Algebra. Vol. 1.* Oxford: Pergamon Press.

Richard D.M. (1990) Gödel Tune: Formal Models in Music Recognition Systems. *Proceedings of the International Computer Music Conference'1990.* Glasgow, 338–340.

Richards W. & Hoffmann D. (1986) Parts of Recognition. In: Pentland A.P. (Ed.) *From Pixels to Predicates.* Norwood, New Jersey: Albex Publishing Corp.

Risset J.-C. (1971) Paradoxes de hauteur: le concept de hauteur sonore n'est pas le même pour tout le monde. *Proceedings of the 7th International Congress of Acoustics.* Budapest, 205–210.

Risset J.-C. (1978) *Paradoxes de hauteur.* Paris: IRCAM Report No. 10.

Roads C. (1980) Artificial Intelligence and Music. *Computer Music Journal,* 4(2), 13–25.

Roads C. (1982) McLeyvier Music Transcriber. *Computer Music Journal,* 6(2), 90–91.

Roads C. (1987) Sinclavier Music Printing. *Computer Music Journal,* 11(2), 77–78.

Rock I. (1983) *The Logic of Perception.* Cambridge, Massachusetts: M.I.T. Press.

Roederer J.G. (1975) *Introduction to Physics and Psychophysics of Music.* New York: Springer.

Rosenthal D. (1988) A Model of the Process of Listening to Simple Rhythms. *Proceedings of the 14th International Computer Music Conference.* Köln: Feedback-Studio-Verlag, 189–197.

Rosenthal D. (1989) A Model of the Process of Listening to Simple Rhythms. *Music Perception,* 6(3), 315–328.

Rosenthal D. (1992) Intelligent Rhythm Tracking. *Proceedings of the International Computer Music Conference'1992.* San Francisco: Computer Music Association, 227–230.

Rossing T.D. (1990) *The Science of Sound.* 2nd ed. Reading, Massachusetts: Addison-Wesley.

Schloss W.A. (1985) *On the Automatic Transcription of Percussive Music. From Acoustical Signal to High Level Analysis.* Stanford: Stanford University, Dep. of Music Report STAN-M-27.

Shepard R.N. (1964) Circularity in Judgements of Relative Pitch. *Journal of the Acoustical Society of America,* 27, 2346–2353.

Shoham Y. (1988) *Reasoning about Change: Time and Causation from the Standpoint of Artificial Intelligence.* Cambridge, Massachusetts: M.I.T. Press.

Skrjabin A. (1960) *Poem for Piano. Op. 32 No. 1. The Text of Author's Performance by Recording on "Velte-Mignon". Transcribed by P.Lobanov.* Moscow: Gosudarstvennoye Muzykalnoye Izdatelstvo. (Russian).

Sundberg J. & Tjernlund P. (1970) A Computer Program for a Notation of Performed Music. Stockholm: Royal Institute of Technology, Report STL–QPSR 2–3/1970, 46–49.

Tanguiane A.S. (1977) *On Parallel Fifths, Doubling, and Orchestration.* Moscow: Moscow State Conservatory. Typescript. (Russian).

Tanguiane A.S. (1987) *Recognition of Chords for Automatic Note Transcription of Polyphonic Music.* Preprint. Moscow: The All-Union Ethnomusicology Commission of the Union of Composers of the USSR, the Computer Center of the USSR Academy of Sciences. (Russian).

Tanguiane A.S. (1988a) Recognition of Chords, Interval Hearing, and Music Theory. In: Alekseev E., Andreeva E., Boroda M., & Tanguiane A. (Eds.) *Quantitative Methods in Ethnomusicology and Music Theory.* Moscow: Soviet Composer, 155–186. (Russian).

Tanguiane A.S. (1988b) An Algorithm of Recognition of Chords. *Proceedings of the 14th International Computer Music Conference.* Cologne: Feedback-Studio-Verlag, 199–210.

Tanguiane A.S. (1989a) Recognition of Chords with the Help of the Model of Interval Hearing. *Doklady Akademii Nauk SSSR*, 308(3), 552–556. (Russian).

Tanguiane A.S. (1989b) A Model of Relativity of Perception and Its Applications to Pattern Recognition in Analysis of Performed Music. *The First International Conference on Music Perception and Cognition, Kyoto, 17–19 October, 1989*, 261–266.

Tanguiane A.S. (1990) A Principle of Correlativity of Perception and Its Applications to Pattern Recognition. *Matematicheskoye Modelirovaniye*, 2(8), 90–111. (Russian).

Tanguiane A.S. (1991a) Recognition of Chords, Perception Correlativity, and Music Theory. *Musikometrika*, 3. Bochum: Brockmeyer, 163–199.

Tanguiane A.S. (1991b) Criterion of Data Complexity in Rhythm Recognition. *Proceedings of the International Computer Music Conference'1991*, Montreal: Faculty of Music, McGill University, 559–562.

Tanguiane A.S. (1992a) Time Determination by Recognizing Generative Rhythmic Patterns. *Musikometrika*, 4. Bochum: Brockmeyer, 83–99.

Tanguiane A.S. (1992b) Artificial Perception and Music Recognition: a Heuristic Approach. *Proceedings of the 10th European Conference on Artificial Intelligence'92, Vienna.* Wiley: Chichester, 169–173.

Tanguiane A.S. (1992c) Artificial Perception and Music Recognition: Theoretical Grounds. *Advances in Artificial Intelligence—Theory and Application. Proceedings of the 6th International Conference on Systems Research Informatics and Cybernetics, Baden-Baden, 17-23 August, 1992, Vol. II.* Windsor, Ontario: the International Institute for Advanced Studies in Systems Research and Cybernetics, 165–170.

Tanguiane A.S. (1993) A Model of Correlative Perception and Its Applications to Music Recognition. *Music Perception.* (Forthcoming).

Teaney D.T., Mourizzi V.L., & Mintzer F.C. (1980) The Tempered Fourier Transform. *Journal of the Acoustical Society of America,* 67(6), 2063–2067.

Terzopoulos D., Witkin A., & Kass M. (1988) Constraints on Deformable Models: Recovering 3D Shape and Nonrigid Motion. *Artificial Intelligence* 36(6), 91–123.

Thibadeau R. (1986) Artificial Perception of Actions. *Cognitive Science,* 10, 117–149.

Ullman S. (1979) *The Interpretation of Visual Motion.* Cambridge, Massachusetts: M.I.T. Press.

Ullman S. (1990a) Aligning Pictorial Descriptions. In: Winston P. & Shellard S. (Eds.) *Artificial Intelligence at MIT: Expanding Frontiers, Vol. 2.* Cambridge, Massachusetts: M.I.T. Press, 344–403.

Ullman S. (1990b) Recovery of 3-D Structure from Motion. In: Winston P. & Shellard S. (eds.) *Artificial Intelligence at MIT: Expanding Frontiers, Vol. 2.* Cambridge, Massachusetts: M.I.T. Press, 404–435.

Ulupinar F. & Navatia R. (1993) Perception of 3-D Surfaces from 2-D Contours. *IEEE Transactions on Pattern Analysis and Machine Intelligence,* 15(1), 3–18.

van der Waerden B.L. (1953) *Modern Algebra. Vol. 1.* New York: Frederick Ungar Publishing Co.

Vercoe B. & Cumming D. (1988) Connection Machine Tracking of Polyphonic Audio. *Proceedings of the 14th International Computer Music Conference, Cologne, September 20-25, 1988.* Cologne: Feedback-Studio-Verlag, 211–218.

Viret J. (1977) Mesure. In: Honnegger M. (Ed.) *Dictionnaire de la Musique. Science de la Musique. Formes, Technique, Instruments.* Paris: Bordas.

Warren R.M. (1982) *Auditory Perception: A New Synthesis.* New York: Pergamon Press.

Wertheimer M. (1923) Untersuchungen zur Lehre von der Gestalt, II. Psychologische Forschung, 4, 301–350. Condensed transl. in: Ellis W.D. *A Source Book of Gestalt Psychology, Selection 5.* New York: Humanities Press 1950. Also in: Beardslee D.C. & Wertheimer M. (Eds.) *Readings in Perception, Selection 8.* Princeton, New Jersey: Van Nostrand Reinhold, 1958.

Widmer G. (1990) The Usefulness of Qualitative Theories of Music Perception. *Proceedings of the International Computer Music Conference'1990.* Glasgow, 341–344.

Widmer G. (1992) Qualitative Perception Modeling and Intelligent Music Learning. *Computer Music Journal,* 16(2).

Wightman F.L. & Kistler D.J. (1989) Headphone Simulation of Free-Field Listening. II: Psychophysical Validation. *Journal of the Acoustical Society of America,* 85, 868–878.

Witkin A.P. (1981) Recovering Surface Shape and Orientation from Texture. *Artificial Intelligence,* 17, 17–45.

Witkin A.P. (1983) Scale-Space Filtering. *Proceedings of the 8th International Joint Conference on Artificial Intelligence, Karlsruhe, West Germany,* 1019–1024.

Witkin A., Kass M., Terzopoulos D., & Barr A. (1990) Linking Perception and Graphics: Modeling with Dynamic Constraints. In: Barlow H., Blackmore C., & Weston-Smith M. (Eds.) *Images and Understanding.* Cambridge: Cambridge University Press.

Witkin A.P. & Tenenbaum J.M. (1983a) What is Perceptual Organization for? *Proceedings of the 8th Joint Conference on Artificial Intelligence'83, Karlsruhe,* 1023–1026.

Witkin A.P. & Tenenbaum J.M. (1983b) On the Role of Structure in Vision. In: Beck J., Hope B. & Rosenfeld A. (Eds.) *Human and Machine Vision.* New York: Academic Press, 481–543.

Wyse L., Carl R., Disher T., & Labriola S. (1985) Sinclavier II Updates. *Computer Music Journal,* 9(4), 81–83.

Xenakis I. (1954) La crise de la musique sérielle. *Graversaner Blätter*, No. 1.

Xenakis I. (1963) *Musiques Formelles*. Paris: Edition Richard-Masse.

Xenakis I. (1971) *Formalized Music*. Bloomington: Indiana University Press.

Zhang Z. & Faugeras O.D. (1990) Tracking and Grouping 3D Line Segments. *Proceedings of the 3rd International Conference on Computer Vision, Osaka.* 577–580.

Zucker S.W., Rosenfeld A. & Davis L.S. (1975) General Purpose Models: Expectations About the Unexpected. *Proceedings of the 4th International Conference on Artificial Intelligence*, 716–721.

Index

Name Index

Subject Index

sum of rhythmic patterns 145
support of spectrum 52
synthesis of new voice 163
tail of spectrum 59
tempered scale 116
tempo 19, 131, 174, 175
tempo curve vi, 132, 138, 148, 174
tempo tracking 5, 132, 179
theorem
    necessary condition for genera-
       tive tone pattern 88
    revealing causality by optimal data
       representation 71
    uniqueness of chord decomposi-
       tion 69
    uniqueness of interval decompo-
       sition 65
thorough bass 160
timbral effect 163
time 131, 147, 175
time events vi, 23, 134
time interval 5, 140
time pattern 175
time resolution 121
tone 47, 177
    representation 47
    spectrum v, 52
      variability with pitch 77
    harmonic 47
      discrete representation 53
    inharmonic 50
    musical 47
tracking simultaneous audio processes
    179
trajectories 18, 36, 179
translation of spectrum 54
transmitting melody from one in-
    strument to another 166
triad, irreducibility 59
trills 170
uniqueness of chord decomposition
    66, 69
unisons 3, 16

vibrato 170
visual scene analysis 8ff, 179
voice crossing 18, 100, 169
voice-leading
    harmonic 170
    melodic 170
    parallel 15, 16
voice overlapping 170
voice separation v, 4–6, 161
voice type, harmonic or inharmonic
    105, 116
voices, parallel 15, 16
weak accentuation 141

Lecture Notes in Artificial Intelligence (LNAI)

Lecture Notes in Computer Science